Interpreting Probability

Interpreting Probability: Controversies and Developments in the Early Twentieth Century is a study of the two main types of probability: the "frequency interpretation," in which a probability is a limiting ratio in a sequence of repeatable events, and the "Bayesian interpretation," in which probability is a mental construct representing uncertainty, and which applies not directly to events but to our knowledge of them.

David Howie sketches the history of both types of probability and investigates how the Bayesian interpretation, despite being adopted at least implicitly by many scientists and statisticians in the eighteenth and nineteenth centuries, was discredited as the basis for a theory of scientific inference during the 1920s and 1930s.

Through analysis of the work of two British scientists, Sir Harold Jeffreys and Sir Ronald Aylmer Fisher, and a close examination of a dispute between them in the early 1930s, Howie argues that a choice between the two interpretations of probability is not forced by pure logic or the mathematics of the situation, but rather depends on the experiences and aims of the individuals involved and their views of the correct form of scientific inquiry.

Interpreting Probability will be read by academicians and students working in the history, philosophy, and sociology of science, particularly in probability and statistics, and by general readers interested in the development of the scientific method.

David Howie holds two doctorates, one in atomic physics from the University of Oxford and one in the history and sociology of science from the University of Pennsylvania. He lives in London.

**Cambridge Studies in Probability,
Induction, and Decision Theory**

General editor: Brian Skyrms

Advisory editors: Ernest W. Adams, Ken Binmore, Jeremy Butterfield,
Persi Diaconis, William L. Harper, John Harsanyi,
Richard C. Jeffrey, James M. Joyce, Wlodek Rabinowicz,
Wolfgang Spohn, Patrick Suppes, Sandy Zabell

Interpreting Probability

Controversies and Developments in the Early Twentieth Century

DAVID HOWIE

CAMBRIDGE
UNIVERSITY PRESS

CAMBRIDGE UNIVERSITY PRESS
Cambridge, New York, Melbourne, Madrid, Cape Town, Singapore, São Paulo

Cambridge University Press
The Edinburgh Building, Cambridge CB2 8RU, UK

Published in the United States of America by Cambridge University Press, New York

www.cambridge.org
Information on this title: www.cambridge.org/9780521812511

First published 2002
This digitally printed version 2007

A catalogue record for this publication is available from the British Library

Library of Congress Cataloguing in Publication data
Howie, David, 1970–
Interpreting probability : controversies and developments in the early
twentieth century / David Howie.
p.   cm. – (Cambridge studies in probability, induction, and decision theory)
Includes bibliographical references and index.
ISBN 0-521-81251-8
1. Probabilities.   2. Bayesian statistical decision theory.   3. Jeffreys, Harold, Sir, 1891–1989
4. Fisher, Ronald Aylmer, Sir, 1890–1962.   I. Title.   II. Series.
QA273.A4 H69   2002   2001052430

ISBN 978-0-521-81251-1 hardback
ISBN 978-0-521-03754-9 paperback

Contents

Acknowledgments

This book grew out of a doctoral dissertation completed in the Department of the History and Sociology of Science at the University of Pennsylvania. I am indebted to my supervisor there, Professor Robert E. Kohler, both for his moral support and for his detailed and constructive criticism of my work. I am also grateful to Professor M. Norton Wise, then at Princeton and the second member of my committee, for his encouragement and advice.

A number of statisticians were willing to answer my queries on the Jeffreys–Fisher debate, and to comment more generally on twentieth-century interpretations of probability. I am grateful here to Professors M.S. Bartlett and I.J. Good. It has been a particular privilege to have corresponded with Professor George Barnard during the last few years. He was also kind enough to read through the entire manuscript, and his insightful comments have helped me avoid many statistical and historical blunders, and have greatly enriched my understanding and feel for the thinking of the period. Professor Paul Forman of the Smithsonian Institution read through an earlier draft, too, as did a philosopher, Dr Matt Cavanagh. Their suggestions have much improved the final text. Thanks, too, for comments and suggestions from Professor John Forrester of Cambridge, and for the support and advice of Dr Elisa Becker of the American Bar Foundation. Ronald Cohen edited the manuscript judiciously.

Special gratitude is due to Lady Jeffreys, for granting me access to her late husband's papers, and for providing me with encouragement and conversation during the weeks I worked in her home researching this book. I deeply regret her passing away before it was completed.

I spent my first two years in Philadelphia as a Fulbright Scholar; thereafter, I was funded by the University of Pennsylvania, first as a teaching assistant, then as a doctoral dissertation fellow. The American Institute of Physics supported a research trip to England in early 1997. I completed my research during the summer of 1998 as a research fellow at the Smithsonian Institution. Finally, I thank the directors and shareholders of Oxford Nanotechnology Plc for their generous support while I finished the manuscript.

1

Introduction

"It is curious how often the most acute and powerful intellects have gone
astray in the calculation of probabilities."

William Stanley Jevons

1.1. THE MEANING OF PROBABILITY

The single term 'probability' can be used in several distinct senses. These
fall into two main groups. A probability can be a limiting ratio in a sequence
of repeatable events. Thus the statement that a coin has a 50% probability of
landing heads is usually taken to mean that approximately half of a series
of tosses will be heads, the ratio becoming ever more exact as the series is
extended. But a probability can also stand for something less tangible: a degree of knowledge or belief. In this case, the probability can apply not just to
sequences, but also to single events. The weather forecaster who predicts rain
tomorrow with a probability of $\frac{1}{2}$ is not referring to a sequence of future days.
He is concerned more to make a reliable forecast for tomorrow than to speculate further ahead; besides, the forecast is based on particular atmospheric
conditions that will never be repeated. Instead, the forecaster is expressing
his confidence of rain, based on all the available information, as a value on
a scale on which 0 and 1 represent absolute certainty of no rain and rain
respectively.

The former is called the frequency interpretation of probability, the latter the epistemic or 'degree of belief' or Bayesian interpretation, after the
Reverend Thomas Bayes, an eighteenth-century writer on probability.[1] A frequency probability is a property of the world. It applies to chance events. A
Bayesian probability, in contrast, is a mental construct that represents uncertainty. It applies not directly to events, but to our knowledge of them, and can

---

[1] I shall use the terms "Bayesian" and "Bayesianism" when discussing the epistemic interpretation, though they were not common until after World War II.

1

thus be used in determinate situations. A Bayesian can speak of the probability of a tossed coin, for example, even if he believes that with precise knowledge of the physical conditions of the toss – the coin's initial position and mass distribution, the force and direction of the flick, the state of the surrounding air – he could predict exactly how it would land.

The difference between the two interpretations is not merely semantic, but reflects different attitudes to what constitutes relevant knowledge. A frequentist would take the probability of throwing a head with a biased coin as some value P, where $P \neq \frac{1}{2}$. This would express his expectation that more heads than tails, or vice versa, would be found in a long sequence of results. A Bayesian, however, would continue to regard the probability of a head as $\frac{1}{2}$, since unless he suspected the direction of the bias, he would have no more reason to expect the next throw to be a head than a tail. The frequentist would estimate a value for P from a number of trial tosses; the Bayesian would revise his initial probability assessment with each successive result. As the number of trial tosses is increased, the values of the two probabilities will tend to converge. But the interpretations of these values remain distinct.

1.2. THE HISTORY OF PROBABILITY

The mathematical theory of probability can be traced back to the mid-seventeenth century. Early probabilists such as Fermat and Pascal focused on games of chance. Here the frequency interpretation is natural: the roll of a die, or the selection of cards from a deck, is a well-defined and repeatable event for which exact sampling ratios can easily be calculated. Yet these men were not just canny gamblers. They intended their theory of probability to account for hazard and uncertainty in everyday life too. This required an inversion of the usual methods. Games of chance involve direct probabilities: from the constitution of a deck of cards, the odds of being dealt a specified hand can be calculated precisely using combinatorial rules. Needed for general situations, however, was the reverse process – working backward from the observation of an event or sequence of events to their generating probabilities, and hence the likely distribution of events in the future.

This problem of 'inverse probability' was addressed by Bayes in a paper published in 1764. His method could be applied to the common model in which draws are made from an urn filled with unknown quantities of differently colored balls. It could be used to calculate the likely proportion of colors in the urn, and hence the probability that the next selected ball will be of a given color, from the information of previous sampling. If the urn is known to contain only black and white balls, for example, and assuming that all possible fractions of black balls in the urn are initially equally likely, and that balls drawn are replaced, then the observation of p black and q white balls from the first $(p + q)$ draws yields the probability that the next ball drawn will be black of $(p + 1)/(p + q + 2)$. The urn models the process of learning from observations. The calculus of inverse probability thus enabled unknown causes to be inferred from known effects.

The method was championed from the late eighteenth century by Laplace as a universal model of rationality. He applied this 'doctrine of chances' widely. In the courtroom, for example, the probability that the accused was guilty could be calculated from the known jury majority, and hence used to assess voting procedures. Probabilists such as Laplace, however, tended not to notice that inverting the equations changed the meaning of probability. The probability of picking a ball of a particular color was a function of the relative proportions in the urn. The probability of guilt was different: it was a quantified judgment in the light of a specified body of knowledge.

Inverse probability began to fall from favor around 1840. Its opponents scoffed at the idea that degrees of rational belief could be measured, and pointed out that since the method could be used only to update beliefs rather than derive them *a priori*, it usually relied on probability evaluations that were arbitrary, or at least unjustified. The mid-nineteenth century also saw a rapid growth in attempts to quantify the human subject, as marked by an eruption of published numerical data touching on all matters of life, particularly crime rates and incidence of disease. The stable ratios observed in these new statistics seemed a firmer base for probability than the vague and imperfect knowledge of the inverse method. Nevertheless, Bayesian methods were often the only way to tackle a problem, and many philosophers and scientists continued through Victorian times to maintain that inverse probability, if only in qualitative form, modeled the process by which hypotheses were inferred from experimental or observational evidence.

As a model of the scientific method, inverse probability reached its nadir around 1930. Yet this eclipse was temporary. Following pioneering work by L.J. Savage, Bruno de Finetti, Dennis Lindley, and I.J. Good in the 1950s, Bayesianism is once again in vogue in statistical, economic, and especially philosophical circles. From the late 1960s, experimental psychologists have typically modeled the human mind as Bayesian, and legal scholars have seriously proposed that jurors be schooled in Bayesian techniques. Bayesian confirmation theory is dominant in contemporary philosophy, and even though most scientists continue to rely on the frequency interpretation for data analysis, the Bayesian calculus was recently used in the observation of the top quark.[2]

This is a study of the two types of probability, and an investigation of how, despite being adopted, at least implicitly, by many scientists and statisticians in the eighteenth and nineteenth centuries, Bayesianism was discredited as a theory of scientific inference during the 1920s and 1930s. I shall focus on two British scientists, Sir Harold Jeffreys (1891–1989) and Sir Ronald Aylmer Fisher (1890–1962). Fisher was a biological and agricultural statistician who specialized in problems of inheritance. In addition to developing new statistical methods, he is credited, with J.B.S. Haldane and Sewall Wright, with synthesizing the biometric and Mendelian approaches into a theory of evolution based on gene frequencies in populations. Jeffreys was a theoretical geophysicist and astrophysicist who also made seminal contributions to the fledgling fields of atmospheric physics and seismology. He worked largely alone to construct sophisticated mathematical models of the Earth and the solar system from the principles of classical mechanics. When tested against the available evidence, such models could be used to investigate the origins and internal structure of the Earth; in this way, Jeffreys inferred in 1926 that the Earth's core was liquid. Jeffreys knew that such conclusions were always uncertain, and that new results could force their revision or abandonment, and tried to account for this with a formal theory of scientific reasoning based on Bayesian probability. Fisher, in contrast, was more concerned with the reduction and evaluation of experimental data than an assessment of how likely a theory was to be true. He regarded Bayesian methods as unfounded in principle and misleading in practice, and worked to replace them with a theory of statistical inference based on frequencies.

The two men developed their theories independently during the 1920s, but clashed publicly in the early 1930s over the theory of errors. This was not simply a mathematical matter. Though eighteenth-century probabilists had been careful to attune the doctrine of chances to the intuitions of men of quality, by the beginning of the twentieth century the probability calculus was taken to define the rational attitude in matters of uncertainty. For both Fisher and Jeffreys, therefore, a theory of probability carried a moral imperative. Not only applicable to analysis, it constituted 'correct' scientific judgment and

---

[2] Bhat, Prosper, and Snyder 1997.

4

fixed the proper role of the scientist. Fisher believed that a scientist should not step beyond the available data. Consequently, he regarded the Bayesian method as inherently unscientific, because initial probability assignments were arbitrary and hence sanctioned individual prejudice. Jeffreys, in contrast, considered science to be simply a formal version of common sense. Though he was concerned that the calculus reflect the general consensus among scientists, he took it as obvious that the expertise and experience of the individual should be a necessary component of a theory of science.

1.4. METHODS AND ARGUMENT

Science has traditionally been presented as a logical and progressive activity in which theoretical models, initially abstracted from facts of observation, are refined or revised with incoming experimental data in order to account for a range of physical phenomena with increasing precision and generality. In the last few decades, however, a number of historians have reacted to this account. The 'scientific method,' they argue, is less a description of usual practice than a fiction used by scientists to bolster confidence in their theories. Science is not an impersonal or faceless activity, but the product of individuals, often with strong political affinities or religious beliefs, and usually eager to win recognition and further their careers.[3] Such concerns can influence research in a number of ways. Though some historical accounts have depicted particular scientific theories as the product of little more than, say, middle-class values or certain ideological commitments,[4] the more persuasive seem to indicate that the effects of such 'external' factors are more subtle. In addition to directing the focus of research, they can shape standards of scientific conduct. Thus various scholars have argued that characteristic approaches to science form around different communities, whether these be marked by distinct experimental apparatus or analytical procedures,[5] or the

[3] An early advocate of this approach was the American sociologist Robert Merton, who from the late 1930s studied the normative structure of science, and how it related both to the wider society and to science as a profession. See the collection of essays, Merton 1983.

[4] MacKenzie 1981 argues that British statisticians' development of new techniques during the nineteenth century was bound up with their middle-class values and commitment to eugenics.

[5] For example, Traweek 1988 relates the experimental practices and career strategies of nuclear physicists to the apparatus of their groups; Buchwald 1994 points to the importance of instrumentation on the interpretation of Hertz's electromagnetic experiments.

wider boundaries associated with disciplines and sub-disciplines,[6] or national groups or even genders.[7] Sometimes, such approaches are borrowed wholesale from more familiar forms of life. Shapin and Schaffer have argued, for example, that the culture of modern science dates to the seventeenth century, and was appropriated from the norms of behavior of the English gentleman.[8] Their study reveals much about how science became authoritative. The credibility of particular results depended on the social status of the individuals who vouched for them. The gentleman could not be doubted without giving him the lie; contrariwise, artisans, or those in employ, could hardly be disinterested. Contrasting with this 'English' style of persuasion is that associated with the nineteenth-century German bureaucratic culture of 'exact sensibility,' which tended to place trust in numerically precise experiments with quantitative error estimations.[9]

Some sociologists have gone further, and argued that in influencing the interpretation of scientific theories, and the criteria by which they are evaluated and tested, such social and cultural factors determine the content of scientific knowledge too.[10] Others have pointed to the holistic character of research, noting that experiments are generally not simple exercises in fact-gathering, but depend on a variety of factors – the equipment, the skills of the experimenter, his theoretical and phenomenological understanding, and so on – that in practice can be reconciled in a number of distinct but consistent ways. Thus an experiment has no single successful outcome, and any particular result must be regarded as local and contingent, in part the choice of the experimentalist, his expectations and presuppositions, and the intellectual and experimental resources available.[11]

---

[6] For example, Heilbron and Seidel 1989 describe the organizational structure that evolved around particle physics; Bromberg 1991 and Forman 1992 relate the development of the laser and maser, respectively, to the utilitarian needs of the sponsoring government and military institutions.

[7] For example, Pestre and Krige 1992 analyze how national work styles influenced the growth of large-scale physics research; Hunt 1991b describes the influence of the 'British' tradition that the aether must be reducible to mechanical principles on those scientists attempting to explicate Maxwell's electromagnetic theory.

[8] Shapin and Schaffer 1985, Shapin 1994.

[9] See the essays in Wise 1995.

[10] Bloor [1976] 1991 is a key text of the 'sociology of scientific knowledge,' a field much influenced by Kuhn's 1962 theory of scientific revolutions. For a survey of the field, see Barnes, Bloor, and Henry 1996.

[11] Hacking 1992 outlines this holistic theory of scientific research. For examples, see Galison's 1987 study of high-energy physics and Pickering's 1989 study of the weak neutral current.

Such approaches have become notorious in recent years, and have been attacked by scientists as undermining their claim that scientific knowledge is reliable.[12] To recognize that science is not 'value free,' however, is not to regard its knowledge as false. Languages are wholly artificial yet still help to explain and manipulate the world. Most historians and sociologists of science are simply attempting to stimulate new directions in their fields rather than questioning the authority or cognitive content of science.

This book is concerned not so much with experimental science as with *theories* of experimental science. The process of theorizing, however, shares with experimentation a creative and open-ended character. Not even a mathematical theory is a straightforward product of logic. Bare equations mean nothing; they must be interpreted through models and analogies, and to be used in science they require rules to govern how and to what they apply.[13] As Wittgenstein indicated, mathematics too is a form of life. My work examines the case of probabilistic theories of scientific inference. The concept of probability has a degree of plasticity, and I will argue that its interpretation and application depends in part on the particular context.

For Fisher and Jeffreys, a strong claim can be made about the relative influence of the various contextual factors. Although with diametrically opposed views of the role of probability in science, they were otherwise very similar. Almost exact contemporaries, both came from respectable but straitened middle class backgrounds to be educated in Cambridge around the same time. They both had wide-ranging scientific interests, and both were early influenced by the polymath Karl Pearson. Indeed, during the 1940s and 1950s, they lived within a minute's walk of each other in north Cambridge. The chief difference stemmed from another similarity. Both were primarily practicing scientists rather than professional statisticians or mathematicians, and each developed his theory of probability in response to specific features of his research discipline. It was these, I shall argue, that was the major influence for each man in selecting his interpretation of probability.

Fisher worked on Mendelian genetics, one of the few wholly probabilistic theories in natural science, and unique in involving a set of clearly-defined outcomes, each of exact probability. A chance mechanism was central to

---

[12] See, e.g., Sokal and Bricmont 1998.

[13] For an example along these lines, see Pickering and Stephanides's 1992 account of 'quaternions' – generalized complex numbers developed by William Hamilton for problems in applied mathematics; also see Warwick 1992, 1993, who has considered Einstein's relativity theory as a broad resource that could be selectively mobilized according to the interests of various groups and individuals.

Mendel's theory, and the outcome of an experiment was to be explained not in terms of uncertain causes or incomplete knowledge, but with reference to the irreducible unpredictability of a definite if initially unknown probability. Fisher considered the results of his breeding and agricultural experiments as random samples from a distribution of fixed probability. This probability had a clear, indeed almost palpable, interpretation as the limiting frequency or ratio of, say, a biological trait appearing in a large ensemble of data.

For Jeffreys, such an interpretation was less natural. His research in geophysics did not fit the Mendelian model. From our knowledge of the Earth, what is the probability that its core is molten? It made no sense to talk of an ensemble of Earths. Instead, the probability represented imperfect knowledge. After all, the core was either molten or not. Geophysics involved much inference from incomplete data, but few of the questions that interested Jeffreys were explicable in terms of repeated sampling. Was it reasonable to believe on the current evidence that the solar system formed by condensation, or that there will be higher rainfall next year than this? Jeffreys decided that all scientific questions were essentially of this uncertain character, and that a theory of scientific inference should therefore be based on probability. Such a probability was not a frequency, however, but clearly a degree of knowledge relative to a given body of data.

Disciplinary concerns entered another way too. Like many of the founders of modern statistical theory, Fisher was a keen proponent of eugenics. This issue was obscured by charged and emotive rhetoric, and Fisher believed that an objective and mathematically rigorous way of dealing with quantitative data could remove individual bias and thus render the lessons of eugenics compelling. Moreover, such a method, if suitably general, could also be used to instill proper scientific standards in new academic disciplines, and prescribe correct experimental procedure. Since he was fashioning statistics as a normative procedure, and the statistician as a generalized scientific expert, with duties not merely of analysis, but in the design of experiments to yield unambiguous answers, Fisher regarded the subjectivity and ambiguity of the Bayesian interpretation with abhorrence. For Jeffreys, on the other hand, matters of public policy were not an issue, and thus strict objectivity not at so great a premium. Fisher's recipes for experimental design were not much help either. One did not experiment on the Earth or the solar system, but relied instead on observations from seismic stations or telescopes. Since this data was often sparse or ambiguous, Fisher's brute statistical recipes were useless. Instead, to tease any genuine physical effects from the scatter and noise, the scientist needed to use his personal expertise and previous experience. Fisher would regard this as a subjective matter, but for Jeffreys it was necessary if any inferences were to be drawn from the available evidence, and was precisely the sort of information that could be encoded in the Bayesian model.

Each interpretation of probability, therefore, seemed suited to a particular sort of inquiry. Even so, perhaps Jeffreys was simply wrong and Fisher right, or vice versa? To meet this objection, I shall examine their dispute in detail. Much of the argument was mathematical, seemingly unpromising territory for the contextual historian. Yet as the 'strong programme' of the Edinburgh school has established – arguably their most constructive product – the deep assumptions and commitments of a research program, though usually tacit, are often unveiled during periods of controversy.[14] The exchange between Fisher and Jeffreys was one of the few occasions in which a prominent frequentist was forced to confront an intelligent and thoughtful opponent rather than a straw man, and indicates that their versions of frequentism and Bayesianism were incompatible but coherent. Neither scheme was perfect for scientific inference, but each had certain advantages – according to Fisher and Jeffreys's different ideas of what counted as genuine and trustworthy science – when compared with the other.

The debate between Fisher and Jeffreys ended inconclusively, each man unyielding and confident in his own methods. Yet in the following decade, inverse probability was almost entirely eclipsed as a valid mode of scientific reasoning. I have argued that the clash between Fisher and Jeffreys was not a consequence of error on one side. Why then was Jeffreys's work unpersuasive? Contemporary scientists tend to be unimpressed with fancy narratives of their discipline; they believe that social or cultural factors intrude to pervert proper scientific behavior rather than constitute it, and retrospectively gloss most scientific disagreements as resulting from inadequate evidence.[15] Yet this reading is not compelling for the clash between the Bayesian and frequentist schools. The mathematics on each side was not in question, any conceptual difficulties with the Bayesian position had been recognized for decades, and there were no new experimental results to resolve the dispute. Instead, it was a particular alignment of the contextual resources used in the 1930s to interpret and evaluate the Bayesian method that ultimately led to its rejection.

One of these was a broad reconception of the limits of knowledge. The epistemic interpretation of probability flourishes in times of determinism. In a mechanistic world there are no real probabilities; instead the concept reflects incomplete knowledge. Thus in statistical physics probabilities provide an approximate description of the motion of vast numbers of atoms in a volume

---

[14] See, for example, Collins 1985.
[15] See Sokal and Bricmont 1998; also Laudan 1990.

of gas. But scientists changed their view of causation during the 1920s and 1930s. According to the 'new physics' – quantum mechanics – the world was inherently stochastic; radioactive decay was random rather than due to some hidden cause. Of course, an indeterministic theory of science does not force a Bayesian interpretation – Jeffreys argued that the very existence of the new physics meant that theories previously held as conclusive were defeasible, and thus by extension that all scientific generalizations should only be asserted with epistemic probability – but the rise of quantum mechanics accustomed physicists to regard chance not as a consequence of imperfect human inquiry but as a feature of nature, and to explain experimental results in terms of probabilities that could be equated with observed frequencies.

Bayesianism suffered for more prosaic reasons too. Fisher's numerical and objective methods held for social scientists the promise of much-needed rigor and credibility. Indeed, so eager were they to adopt his very particular model of scientific experimentation that they reconfigured some fields, such as psychology and to some extent clinical medicine, to conform with it. And if social scientists were keen to buy objective methods, statisticians were keen to sell. Their rejection of Bayesianism can be seen as part of a strategy of professionalization: by presenting theirs as a general and objective method of analysis, statisticians were justifying their discipline as distinct from any particular field, and thus a candidate for separate university departments, journals, and so on.

The precise combination and influence of such factors differed between disciplines. Some statisticians regarded Fisher's frequentism as no less conceptually vulnerable than Jeffreys's Bayesianism, and rejected both accounts of inference in favor of the severe frequentism of the decision-theory approach developed during the 1930s by Jerzy Neyman and Egon Pearson. Some scientists also argued that a theory of probabilistic inference was not necessary for science. The astronomer Sir Arthur Eddington, the most flamboyant popularizer of Einstein's Theory of Relativity in Britain, asserted loudly that particular general laws must be true on purely *a priori* considerations, and cited Einstein's theory as a case in point. Probability was to be relegated to data analysis. In Germany, the philosopher-physicist Richard von Mises was developing his own frequentist interpretation of probability, but also regarded the concept as having only a restricted place in scientific inquiry. Most mathematicians, though still loosely frequentist, preferred from 1933 to follow the Russian Andrei Kolmogorov with an axiomatic definition of probability based on measure theory. Philosophers persisted with a broadly epistemic probability, but turned for its exposition not to Jeffreys but to John Maynard Keynes's logical interpretation of 1921, or occasionally to that of the mathematician

Frank Ramsey, who in 1926 had sketched the probability calculus as rules for the consistent assignment of degrees of belief, in practice measured by betting odds.

As the concept of probability fragmented, so Fisher and Jeffreys, having battled from opposite sides of a seemingly sharp boundary, found themselves jostled ever closer. In his later papers, Fisher adopted subjectivist language to defend the pragmatic aspects of his own theory against the Neyman–Pearson school. Jeffreys too saw Fisher's approach as more level headed than the alternative versions of non-scientists; soon after their exchange, Jeffreys wrote that his and Fisher's methods almost invariably led to the same conclusions, and where they differed, both were doubtful. Indeed, the Bayesian revival of the 1950s was based not so much on Jeffreys's work as on a more extreme version – personalism – developed, also during the 1930s, by the Italian Bruno de Finetti. De Finetti rejected Jeffreys's attempts to produce standard prior probabilities and his definition of a probability as a degree of knowledge or rational belief. A probability no longer related a hypothesis to a body of data, as Jeffreys maintained, but represented subjective belief, and consequently one set of prior probabilities was as good as any other.

1.5. SYNOPSIS AND AIMS

The study starts in Chapter 2 with an account of the history of probability from the mid-seventeenth to the end of the nineteenth centuries, stressing the relative positions of the epistemic and frequency interpretations. For most of the eighteenth century they were in uneasy mixture, as the classical probabilists applied a single calculus to games of chance, annuities, the combination of observations, and rational action in the face of uncertainty. The distinction was only clearly drawn around the beginning of the nineteenth century. By 1850, the success of astronomical error theory, and the rise of social statistics – itself largely driven by bureaucrats to simplify taxation and administration – made the frequency interpretation more natural. The mathematics of the theory, however, remained intimately connected with particular implementations. Many new statistical techniques were developed by the biometric school expressly to attack particular problems of heredity and evolution, for example. (Ghosts of the biometricians' work survive in the current statistical terminology of regression and correlation.) Yet away from the reduction and analysis of data, many scientists continued to regard inverse probability, and the associated epistemic interpretation, as the model for uncertain inference.

Chapters 3 and 4 examine frequentist and Bayesian probability in more detail. I concentrate on the versions due to Fisher and Jeffreys, and argue that in each case, the choice of interpretation and its subsequent development evolved with a particular experience of scientific practice. Though the treatment is roughly symmetric, these chapters are of unequal length. With his major contributions both to statistics and the modern theory of evolution, Fisher is already recognized as a major figure in the history of science, and has generated a large secondary literature, including an excellent biography by his daughter, Joan Fisher Box. Jeffreys is less well known, yet deserves attention not only because of his contributions to geophysics and probability theory, but because his work style – the solitary theorist introducing rigorous mathematics to a range of fields previously associated with the vaguer methods of the enthusiast or dilettante – is now rare in a profession characterized by rigid boundaries between sub-disciplines. Moreover, though the frequency interpretation of probability and statistical model of experimental design is familiar from standard courses in statistics, the epistemic interpretation, despite the current revival of Bayesian probability theory, is still little known. Thus in addition to making the historiographic points – that even mathematical theories have no unique meaning, but are 'enculturated,' and therefore that a choice between them is not forced by content, but depends also on the context – my work is intended to serve as an introduction both to Bayesian probability theory and to the scientific career of Harold Jeffreys.

The backbone of the study, Chapter 5, concerns the Fisher–Jeffreys debate. By directly comparing frequentist and Bayesian methods, my discussion is intended to continue the explanations begun in previous chapters. It also makes the point that the difference between the approaches of Fisher and Jeffreys stemmed not from mere error or misunderstanding, but from distinct views of the character and purpose of scientific inquiry.

In Chapter 6 I move from Fisher and Jeffreys to ask why the Bayesian position was generally deserted in the years up to the World War II. The discipline of statistics, like economics, effaces its history. Perhaps anxious to encourage the appearance of objectivity, statistics textbooks – especially primers for researchers in other fields – make little mention of controversies in the development of the subject. Thus Bayesianism is usually suppressed, and an inconsistent mixture of frequentist methods, taken from both the Fisherian and Neyman–Pearson schools, is presented as a harmonious 'statistics.' This section of the study is intended as a first step at correction. It outlines the development of probabilistic thought during the 1930s across the disciplines of statistics, the social sciences, the physical and biological sciences, mathematics, and philosophy.

Chapter 6 is not, however, intended to be comprehensive. I do little to distinguish the several variations on the frequentist theme that either emerged or were developed during the 1930s. These include the theories associated with the names of Neyman and Pearson, von Mises, Reichenbach, Popper, Kolmogorov, and Koopman. Each enjoyed a period of support in at least one field, but apart from the Neyman–Pearson theory, none exerted much general influence in the English-speaking scientific world. It is true that von Mises's approach had an impact, and that Kolmogorov's formulation, in terms of measure theory, laid the foundation for a fully consistent theory of probability. But the influence of both developments was largely limited to pure mathematicians rather than practicing scientists. Neither von Mises nor Kolmogorov was widely cited in statistics texts or debates of the Royal Statistical Society during the period I consider. Fisher might have been expected to welcome such frequentist contributions, yet rarely referred to their work in his papers, even when discussing the theoretical and mathematical foundations of the subject. He considered von Mises's theory too restrictive to be of much use in scientific research, and Kolmogorov's too mathematical.[16] Hence my discussions of von Mises and Kolmogorov in Chapter 6 are brief. For similar reasons, I do not consider the subjective theory of de Finetti here at all. Though he started work on probability in the late 1920s, and developed his theory in a number of important papers through the 1930s, de Finetti was an isolated figure until the 1950s even in Italy. Jeffreys saw none of his work, and had not even heard of him until around 1983.[17] The main aim of Chapter 6 is not to distinguish these various sub-categories, but to chart the growing ascendancy of frequentistism over Bayesianism during the 1930s.

The final Chapter 7 is in the form of a brief epilogue followed by conclusions, and shows how the methodological position outlined in this introduction might be applied to the post-war history of probability.

Appendix 1 is a short bibliographical essay of sources used in Chapter 2. Appendix 2 presents the mathematical case for inverse probability as a model of scientific inference. Appendix 3 lists the abbreviations used in the footnotes.

---

[16] In a letter of 1946 Fisher ironically reported his discovery that some of his arguments had apparently been justified by "a certain, I believe very learned, Russian named Kolmogoroff"; he was still claiming to be unfamiliar with Kolmogorov's axioms around 1950.

[17] For more on von Mises, Kolmogorov, and de Finetti, see von Plato 1994.

2

Probability up to the Twentieth Century

2.1. INTRODUCTION

The calculus of probability is conventionally dated from July 1654, when Blaise Pascal wrote Pierre de Fermat with a question raised by a friend, the Chevalier de Méré, concerning a dice game. The subsequent correspondence, ranging widely over gambling problems, was not the first time that games of chance had been addressed mathematically. Muslim and Jewish mathematicians in the twelfth and thirteenth centuries had calculated combinatorial rules, and Renaissance scholars in the fifteenth and sixteenth centuries had analyzed card games and dice throws in terms of the number of ways of reaching each possible outcome.[1] Yet the two savants were the first to treat the subject in a consistent and unified way. The 'problem of points' posed by the Chevalier – concerning the division of stakes between players when a game is interrupted before reaching conclusion – had resisted attempts at solution from the fourteenth century. Pascal and Fermat gave the first correct general solution to this and other problems, and developed new mathematical techniques to calculate the odds in a number of card and dice games.[2]

---

[1] A book on cards and dice by the colorful figure of Girolamo Cardano, astrologer, philosopher, charlatan, and heretic, was finished around 1564, though not published until 1663, eighty-seven years after his death; Galileo wrote a short work on dice games early in the seventeenth century at the instigation of a group of Florentine noblemen.

[2] The Chevalier de Méré, an experienced and intuitive gamesman, had also asked why he tended to lose bets that a double six would be shaken within twenty-four throws of a pair of dice, but win when the number was increased to twenty-five. After all, four throws of a die would more often than not result in at least one six, so if each of the six faces of the second die were equally probable, twenty-four throws of the pair should be enough to win an even bet of obtaining at least one double-six. As Pascal and Fermat explained, however, though the Chevalier's observations were correct, his reasoning was not. The bet changes from being favorable to unfavorable between twenty-four and twenty-five throws because the chance of success, $1-(1-1/36)^n$, is 0.4914 when n is 24 and 0.5055 when n is 25. (The Chevalier must have been a man of keen intuition, for to have reached this judgment on experience alone, even to the level of 80% certainty, would have required him to have watched some 85,000 games. See Holland 1991.)

Following Pascal and Fermat but working largely independently, Christian Huygens derived similar results in his 1657 essay, *De Ratiociniis in Ludo Aleae*. Pierre Rémond de Montmort calculated the expected gains in several complex card and dice games in a publication of 1708. Jacob Bernoulli worked on the subject from 1685 to his death in 1705; his *Ars Conjectandi*, the first systematic and analytically rigorous conspectus of the probability calculus, was published posthumously by his nephew Nicholas in 1713. Abraham De Moivre extended the mathematics in 1718, incorporating improved combinatorial relations, new techniques of infinite series expansions and approximate distributions, and an early form of limit theory into his gambler's bible, the *Doctrine of Chances*.

De Moivre also gave a definition of probability. Previously, the concept had been left obscure. The early writers tended to base their analysis on the fairness or otherwise of a game; indeed, neither Pascal nor Fermat had used the word 'probability' in their letters. Huygens had assessed the value of a gamble by introducing the concept of mathematical expectation, and Bernoulli had decomposed this expectation into a combination of probability of success and winnings if successful. But it was De Moivre who clearly stated that it was "the comparative magnitude of the number of chances to happen, in respect of the whole number of chances to either happen or to fail, which is the true measure of probability."

2.2. EARLY APPLICATIONS OF THE PROBABILITY CALCULUS

Games of chance, in which the possible outcomes were precisely specified and easily countable, were an obvious application of the new mathematical calculus. So was another form of gambling: the sale of annuities and insurance. These financial tools were well established by the mid-seventeenth century. Maritime insurance, for instance, dates back to the 1400s in seafaring nations such as England and Holland. Yet insurers had tended to ignore the occasional theoretical treatment of their subject – such as the Londoner Richard Witt's publication on annuities of 1613 – and instead, like money lenders of the day, offered fixed rates of commission across the board.

This situation started to change with the appearance of empirical data. John Graunt published the first life tables in 1662. Based on the causes of death given by the London Bills of Mortality – instituted in 1562 as a warning of plague epidemics – and with the simplifying assumption that the chance of death remained constant after the age of six, the *Natural and Political Observations* estimated the fraction of a population that would survive to successive ages. Graunt intended his tract to guide the 'art of government.'[3] His estimate of the population of London – at one-third of a million, much lower than the number usually assumed – was of immediate interest. Philosophers and statesmen such as Hume, Rousseau, and Montesquieu regarded such elusive population figures as an index of prosperity, and thus of the power of a nation's ruler and the benevolence of its system of government, or the importance and political value of a new dominion.[4] But early probabilists such as Huygens realized that this population data could be combined with the new probability calculus to yield improved values for annuity and insurance premiums. The Dutchman Johann De Witt used Huygens's formulae in 1671 to calculate annuity rates; the following year, the astronomer Edmund Halley – who went on in 1693 to publish the first complete empirical mortality tables – showed that annuities sold by the British government, then returning 6% on a seven-year lifetime, were undervalued. A number of tables and manuals on the subject started to appear from the beginning of the eighteenth century, including De Moivre's theoretical account of 1725, *Annuities Upon Lives*.

Astronomy, and in particular the problem of the combination of observations, provided the other main application of early probability theory. A planet or star has a certain position in the sky, yet changeable atmospheric conditions, together with inaccurate instruments and imperfect eyesight, will always introduce a scatter into a series of measurements. Astronomers had long debated what to do about this. Kepler was typical in favoring the mean value, which he justified with an appeal to fairness. Galileo, in contrast, borrowing the legal argument that the most reliable story is that supported by the most witnesses, often chose the mode.[5]

Such guidelines for the combination of measurements had evolved by 1700. Roger Cotes, an assistant of Newton, recommended an average weighted according to the errors of single observations, for instance.[6] Yet the need for a standard method grew acute as scientists communicated more freely and their data became increasingly precise and abundant. From the middle of the

---

[3] Malthus's study of 1798 is the best-known example of such 'political arithmetic.'
[4] See Rusnock 1995.
[5] Galileo discussed observational errors in detail in his 1632 discussion of the nova of 1572. (This was a charged topic since Galileo's eventual placement of the nova beyond the Moon was contrary to the prevalent Aristotelian world-view of perfection in the superlunar crystalline spheres.) He recognized that discrepancies between observations of the nova and its real position were more likely to be small than large, and that overestimates and underestimates were equally likely.
[6] Stigler doubts that Cotes's proposal, published in 1722, six years after his death, had much influence, noting that its earliest citation was by Laplace in 1812 (Stigler 1986, p. 16).

eighteenth century, astronomers started to turn to the probability calculus for a more systematic approach to the combination of observations. The face of the Moon visible from the Earth is not fixed but subject to a variation, and in a study of this 'libration' published in 1750, the astronomer Tobias Mayer obtained mean values, with an assessment of their accuracy, for three unknowns from a set of twenty-seven coupled equations. Such work was often undertaken to test Newton's gravitational theory and thus lead to improved methods of navigation. Also important in this regard was the precise shape of the Earth, which due to its rotation was thought to bulge at the equator. Testing this thesis in 1755, Roger Boscovitch turned to probabilistic principles to combine length measurements of short arcs of the Earth's surface taken at different latitudes. In the same year, the mathematician Thomas Simpson applied De Moivre's methods to astronomical observations. (He had previously borrowed from De Moivre for short treatises on games of chance and annuities.) Simpson concentrated on the errors rather than the observations, and took the important step of regarding them, for repeated measurements in identical circumstances, as having a fixed and continuous distribution of probability. Using De Moivre's series expansions, he compared the probability that the mean of a number of errors lies within a specified range with the probability associated with a single error. His conclusion – that means were to be preferred over single measurements – was hardly original, but the probabilistic reasoning was new. The more influential Johann Lambert made a similar advance in Germany around 1760 in supposing that instruments have their own error distributions; additionally, he made an empirical study of the resolution of the eye.

2.3. RESISTANCE TO THE CALCULATION OF UNCERTAINTY

Despite these theoretical successes, probability was slow in becoming established. The concept itself was still esoteric: gamblers were more interested in the fair price of a game, or their expected profit in playing, than in a separate and more nebulous measure of the likelihood of a given outcome; the emphasis among insurers was likewise on an individual's life expectancy rather than the abstract probability of death.[7] On a more fundamental level, many thinkers

---

[7] Hacking considers that the first printed appearance of the term 'probability' in the modern sense, as a numerical measure of uncertainty, occurs in the 1662 book *La Logique, ou L'Art de Penser*, the relevant sections written by the Jansenist theologian Antoine Arnauld. Arnauld recognized the connection between probability and rational behavior. Discussing the danger of lightning strikes, he wrote: "Fear of harm ought to be proportional not merely

were also suspicious of the very idea that chance events could be measured or predicted. Indeed, a 'calculus of probability' was commonly viewed as contradictory. Chance events, by their nature, were uncertain and thus unknowable. A matter of fate or control by the gods, they were not susceptible to precise mathematical treatment.

The rise of natural philosophy during the seventeenth and eighteenth centuries made numerical treatments of physical phenomena common. And by the 1750s, respected mathematicians such as Leonard Euler had tackled increasingly advanced matters in probability, such as duration-of-play problems. Yet despite their direct opportunities to profit, gamblers remained wary of the calculus. The story of the Chevalier de Méré is illuminating. It had been known for centuries that the chance of a given outcome in a dice game was connected with the number of ways of reaching that outcome, yet the Chevalier, in common with scores of other charm-carrying gamblers, was prepared to wager vast sums on little more than intuition and the irrational but unshakable belief that he was on a lucky streak or otherwise fated to win.

Level-headed insurers were also reluctant to embrace the theory. In part, they shared the vulgar belief that a shipwreck or accidental death was a matter for divine will rather than calculation. They were also leery of adopting a mathematics associated with the louche and rather frivolous gambling illustrations that bulked large in the textbooks. Further, the booming insurance industry was too profitable to encourage much attention to dreary mathematical minutiae. But as Theodore Porter has shown, insurers of the time were also strongly opposed to any attempts to quantify their discipline. The broad experience and fine judgment of the professional, they argued, could not be replaced by equations and numerical prescriptions. The decision, on the basis of health, age, and occupation, whether or not a particular individual was appropriate for coverage required nice judgment that could not be supplanted by the blind average of a demographic table.[8] In consequence, the scholarly treatment of insurance and annuities remained relatively flimsy throughout the eighteenth century. There was no consensus whether mean or median ages should be used to reckon annuity rates, for example, and even the more mathematically sophisticated workers supplemented their equations with sparse

to the gravity of the harm, but also to the probability of the event, and since there is scarcely any kind of death more rare than death by thunderstorm, there is hardly any which ought to occasion less fear." See Hacking 1975, pp. 25, 73–9; quote from p. 77.

[8] Daston has noted that life insurance policies, unlike annuities, tended in eighteenth-century England to be taken out on third parties. Essentially bets on the lives of specified celebrities, such as a politician or the Pope, these policies were not as susceptible to probabilistic or empirical study. See Daston 1988, pp. 122–3, 163–8.

demographic data still taken from parish birth and death records. They justi-fied their unrealistic and simplistic assumptions concerning death-rates and infant mortality with glib appeals to natural theology or the 'simplicity of nature.' Thus De Witt's scheme was never put into practice, and the Equi-table Insurance Company, which started to base its premiums on probabilistic calculations around 1762, remained for many years the exception in an in-dustry accustomed to charging flat rates for any granted policy.[9] Despite driving the mathematical development of the theory from the mid-eighteenth century, the insurance market only came to embrace the probability calcu-lus wholeheartedly from the beginning of the nineteenth.[10] In the same way, some astronomers resisted the probabilistic treatment of errors on the grounds that such objective procedures detracted from their individual expertise. This attitude was most noticeable with the occasional appearance of 'outliers,' or incongruous data-points. Experimenters argued that having spent many months building apparatus and making observations, they were in the best position to judge which measurements could be trusted and which simply ignored.

2.4. THE DOCTRINE OF CHANCES

By the end of the eighteenth century, however, probability theory had a more controversial application. In his *Ars Conjectandi* of 1713, Jacob Bernoulli had discussed not only gambling problems but mathematical uncertainty in everyday life. Just as the calculus of probability could be applied to the throw of a die or a hand of cards, he argued, so it could be extended to other forms of uncertainty, such as illness or the weather. Bernoulli maintained that a belief in a cause or theory varied in exactly the same way as the observed ratio of favorable to unfavorable cases. ("For even the most stupid of men ... is convinced that the more observations have been made, the less danger there is of wandering from one's goal.") A matter of knowledge could therefore be likened to a gamble, with evidence or argument worth a fraction of the total stakes depending on its strength, and certainty to the entire pot.

But how could such a probability – a degree of certainty – be measured? As a bridge from games of chance to what he called 'civil' phenomena, Bernoulli introduced the model of an urn filled with balls, each either black

---

[9] Richard Price published on annuities at the instigation of the Equitable in 1771. His widely quoted figures were based on data of Northampton, which subsequently turned out to be unrepresentative – they significantly underestimated life expectancy – meaning many annuities of the following years were under-priced.

[10] See Daston 1987.

or white. The ratio of the number of balls of each color to the total number represented its probability, in the same way as did ratios of equally-possible cases in gambling games. With the risk of a disease recast as a drawing from an urn filled with balls in proportion to the frequency of occurrence as given by medical records, the chances of infection in a given sample could be calculated by combinatorial analysis, and thus the relative efficacy of various treatments tested. The result of each draw was of course a matter of chance. But Bernoulli showed, with what Poisson later christened the 'weak law of large numbers,' that given a large enough number of trial draws, the relative frequency of colors obtained could be taken with 'moral certainty' to equal the actual proportions in the urn.[11]

A theory of knowledge is a more noble thing than an analysis of card games, and probabilists welcomed the opportunity to apply their equations to civil and moral matters. De Moivre used the 'analysis of hazards' widely in economic and political cases. Philosophers also started to take notice. Leibniz, who was fascinated by demographic data and had written in 1666 an interesting if not invariably correct pamphlet on the probability of dice permutations,[12] corresponded extensively with Jacob Bernoulli on the new concept. Following the Reformation, philosophers had started to reject the old religious certainties and instead adopt forms of skepticism. Enlightenment thinkers such as John Locke promoted the idea of a limited knowledge based on considered judgment and the evidence of the senses. Bernoulli's work seemed relevant. Since a reasonable man would act to maximize his expectation of gain, the classical calculus of probabilities – often called in this context the 'art of conjecture' or even the 'reasonable calculus' – could be used to model rational decisions and hence serve as a guide to conduct. Pascal's wager of 1669 is a notorious example: belief in God is secured not blindly or through revelation, but after a cool appraisal of the mathematics. And David Hume argued from probability that no amount of testimony could be sufficient to render plausible historical accounts of religious miracles. As Ian Hacking writes, "Probability was the shield by which a rational man might guard himself against Enthusiasm."[13] The language of probability also started to enter scientific discussions, with theories no longer being regarded as certain or proved, but merely as improbable or probable. Thus Dalton in 1803 defended the caloric theory of heat by declaring it to be the 'most probable opinion' given the evidence.

---

[11] Streamlined versions of Jacob Bernoulli's law, with tighter bounds on the degree of approximation, were produced by his nephew, Nicholas, and subsequently by Poisson.

[12] The probability of throwing eleven with two dice is not the same as throwing twelve.

[13] Hacking 1990, p. 89.

An important illustration of the link between probability for games and as a guide to rational action arose early in the eighteenth century. In a letter to Pierre de Montmort, Nicholas Bernoulli asked the fair price to participate in a game in which a coin is tossed until a head shows, the prize starting at one dollar if the first toss is a head and doubling after each successive tail. The chance of a lengthy run halves with each extra tail, but the pay out on such a sequence is proportionately large. The mathematical expectation, and hence entry fee for a fair game, is apparently infinite. Yet as Bernoulli pointed out, a reasonable man would not pay any more than a few dollars to take part.

This was regarded as considerably more than an amusing problem in mathematics. Known as the St. Petersburg problem, Bernoulli's question provoked an energetic debate that involved most of the prominent mathematicians and philosophers of the eighteenth century. Why all the fuss? The answer is suggested by the words 'reasonable man.' The probability calculus was regarded not an approximation, but as a direct description of rational judgment. Hence any discrepancy between the numbers and sensible action could in principle cast doubt on the art of conjecture as a guide to action. In practice, the situation was more fluid. Nowadays, the theory of probabilistic induction is often held up as a formalization of scientific method, the way a scientist ought to think. In the eighteenth century the position was reversed. 'Reasonableness' was taken as a characteristic innate to men of quality – which is why early probabilists like Pascal and Fermat saw no circularity in basing a definition of probability of the intuitive notion of fairness – and instead it was the equations of the embryonic calculus that needed tweaking when they conflicted with good sense. Responses to the St. Petersburg problem thus constituted attempts to reconcile the mathematics with the behavior of men of quality.[14] For example, in 1738 Daniel Bernoulli introduced the idea of 'utility.' The value of money, he argued, is not constant for a rational man. Unless you are particularly greedy or profligate, an extra dollar means less if you are rich than if you are poor. Illustrating the point with examples from the world of commerce, Bernoulli suggested that the value of money went as the logarithm of the amount held, and was thus able to produce a non-infinite entry fee to play the St. Petersburg game.[15] As a model of rational judgment, therefore,

---

[14] Jorland 1987 reviews attempts to solve the St. Petersburg problem between 1713 and 1937.

[15] This early attempt to quantify value avoided an infinite expectation but hardly closed down the debate. The traditionalists, such as Nicholas Bernoulli, took any extension of probability from fair games as inherently illegitimate. Others pointed out that unless the utility function of monetary units is bounded, a game could always be concocted with a steep enough pay-out to ensure that an infinite expectation would still arise. (Note, however, that the expectation of the game is infinite only if the player offering the bet has

the probability calculus had a certain elasticity. Even so, it carried a persuasive force. Proponents of the numerical art of conjecture expected that when properly calibrated against the intuitions of men of quality, it would force a consensus between disputatious parties and could serve as a guide to action for the masses. It was even suggested that in some cases a machine could be rigged up to deliver judgment mechanically.

This prescriptive element was most plain in the application of probability theory to the weighing of evidence and testimony. Though qualitative degrees of certainty were long established in law, it was not until the end of the seventeenth century that lawyers started to express reliability in numerical form. But it was not obvious how to combine these values. How much more reliable were two witnesses to an event than one? What if they had been together at the time, or conferred afterwards? By how much more could a man be trusted than a woman? A nobleman than a peasant? Some early rules had been produced by 1700. The English divine George Hooper described one scheme in 1689, and ten years later John Craig produced another, with an elaborate Newtonian metaphor that likened mechanical forces to the force of argument. But it was Jacob Bernoulli in *Ars Conjectandi* who first applied the classical calculus, using his balls-and-urn model to generalize Hooper's rules.

Though Montmort in 1713 criticized the application as unrealistic, questions of legal testimony became one of the staples of the probability calculus. Both Nicholas Bernoulli and De Moivre addressed the issue. The application had its heyday with the judicial reforms in France around the time of the Revolution. The Marquis de Condorcet, a brilliant mathematician and one of the first to treat a probability as a degree of credibility, regarded the English model of unanimous decision as inefficient. Instead, he declared that jurors should each be able to form their own judgment and a decision taken by majority vote. Condorcet used probability theory in 1785 to calculate optimum jury sizes and voting procedures. To force the problem into the shape required by the balls-and-urn model required a number of dubious assumptions, such as that each juror acts independently, and that each has an equal probability of correctly assessing the accused as innocent or guilty. Yet the probability calculus was held in such high regard that Condorcet's recommendations – including a jury thirty strong – were briefly put into practice.

the ability to produce arbitrarily large amounts of wealth. After all, the chance of winning more than N dollars in a single sequence drops to zero rather than $\frac{1}{2}^N$ if the size of the bank is limited to N units. This is not a trivial point. Since the doubling effect raises the pot rapidly, the 'fair price' when paying against a total possible pot of 1 million dollars is a reasonable-looking 21 dollars; it only climbs to 31 dollars for a billion-dollar pot. See Epstein 1977, pp. 90–4.)

The classical probabilists worked on the probabilities of specific outcomes in well-defined situations, such as being dealt a certain hand from a pack of cards of known composition, or an innocent man being found guilty by a set number of jurors each with a fixed probability of judging correctly. Starting with Jacob Bernoulli, they started to investigate the opposite direction, of generating probabilities from observed outcomes. Knowing the composition of balls in the urn, combinatorial analysis gives the probability of drawing a given sample. But how can one move from the evidence of a number of draws to the composition of balls in the urn?

This question of 'inverse probability' was glossed as a relation between cause and effect. The proportion of colored balls in a sample was caused by the ratio in the urn, with the limit theorem guaranteeing that these causes would emerge from chance fluctuations in the long run. Since probability and knowledge were thought to be connected, this was a matter of inductive reasoning. Having picked a sample of balls from the urn, one could calculate its likely composition and thus the probability that the next ball drawn would be of a given color. Inverse probability could be used to infer unknown causes from known effects.

The inversion of the usual methods exercised probabilists through the eighteenth century but proved frustratingly elusive. Bernoulli, working with his urn model and the binomial distribution, was not able to produce anything satisfactory. The solution was finally discovered by Thomas Bayes, a non-conformist minister from Tunbridge Wells. Though with few publications to his name – a defense of Newton's theory of fluxions and a paper on divergent series – Bayes was an adroit mathematician. He had studied with De Moivre, and been elected to the Royal Society in 1742. Bayes started to work on inverse probability around 1749, possibly as a direct response to Hume's explicit challenge for a solution to the riddle of induction, thrown down in the *Enquiry Concerning Human Understanding* of the previous year.[16] Although Bayes had not published on the subject by the time of his death in 1761, his friend, the Unitarian minister and political radical Richard Price, found among his papers an "Essay towards solving a problem in the doctrine of chances." This he communicated to the Royal Society in 1763. Describing how the probability of a given cause changes as events accumulate, the essay appeared in the *Philosophical Transactions of the Royal Society* the following year.

---

[16] See Zabell 1989a, pp. 290–3; Dale 1991, pp. 88–9.

Though subjected to Talmudic scrutiny in recent years, Bayes's paper had little impact at the time. It was instead the Marquis de Laplace who popularized the method of inverse probability. The most celebrated scientist in France, Laplace had been introduced to the theory of probability, possibly by Condorcet, in connection with his astronomical studies. This was not a coincidence. The concept of inversion is easier to grasp from within an astronomical framework. An observation of a star is made up of its true position together with an error component. Following the innovations of Simpson and Lambert, it was possible to consider not just the observations, but the errors themselves as forming a random distribution. Thus from a given set of observations, the probability distribution for the unknown position could be produced directly. The errors of observation constitute the random contribution in both directions.

Laplace first discussed the probability of causes, apparently unaware of Bayes's work, in a paper of 1774. Using Bernoulli's model, he showed that if from an infinitely large urn the first $(p + q)$ draws had yielded p black and q white balls, the probability of a further black ball on the $(p + q + 1)^{th}$ draw, assuming that all possible fractions of black balls in the urn were initially equally likely, is $(p + 1)/(p + q + 2)$. So if every one of m drawn balls is black, the probability is $(m + 1)/(m + 2)$ that the next one will be so too. (This formula was named the 'Rule of Succession' by John Venn in 1866.[17])

Laplace's analysis can be regarded as a probabilistic treatment of enumerative induction: if instead of color, some binary property of an object or process is considered, repeated independent trials will lead to a quantitative prediction of future observations. Laplace gave an illustration. What is the probability that the sun will rise tomorrow, assuming the probability of rising is fixed but initially unknown, and given that it has risen each day without fail in 5,000 years of recorded history? The example was not a casual choice. David Hume originally cited the rising of the sun in 1739 to argue that some things, although not necessary in the logical sense, could nevertheless be known with certainty. Several writers had subsequently used it to illustrate their attempts at a probabilistic theory of induction. Richard Price, communicator of Bayes's essay and opponent of Hume's empiricist philosophy, interpreted the probability that the sun will rise tomorrow as the odds of its having successively risen for a million days – roughly 3,000 years – given an even chance each time. The naturalist the Comte de Buffon tackled the

---

[17] Dale notes that with charity one can see a precursor of Laplace's expression in the work of Moses Mendelssohn (1729–1786), the grandfather of the composer Felix (Dale 1991, pp. 52–3).

problem in 1777 in the same way. Laplace's solution, in contrast, enabled the assessment of probability to be updated in the light of new evidence. With each appearance of the sun, the probability of its rising tomorrow increases, asymptotically approaching certainty. Thus did the Rule of Succession justify induction.

In his paper of 1774, Laplace used his method to solve the problem of points, before moving to more practical matters and showing that for three measurements at least, the mean was a better average than the median. In 1778 he wrote on the binomial case of male-to-female birth ratios, and subsequently worked from distributions of samples to estimates of population sizes. He was also able to invert a famous problem in astronomy and find the probability that some cause, rather than chance, had resulted in orbits of the planets being so nearly coplanar. (Daniel Bernoulli had tackled this problem in 1734, but had used direct probabilities, inferring a divine cause from his calculations that the coplanar orbits were otherwise extremely unlikely.) Joseph Louis Lagrange, who around 1770 had extended the work of Simpson and Lambert with a probabilistic assessment of various error distributions, was one of the first to use inverse methods. Condorcet, who though a friend of Price learned of inverse probability from Laplace rather than Bayes, also took to the method eagerly as a tool for his 'social mathematics.' By his death in 1794 he had applied the formula almost indiscriminately to moral data, using inverse probabilistic arguments to argue for educational reform, for example, and to value property according to ownership rights. Yet for most problems, inverse probability was too awkward to use. Extending Mayer's work in a study of Jupiter and Saturn, Laplace found that for real astronomical observations, assuming even a simple form of error distribution made the mathematics practically intractable. He continued to be flummoxed through the 1780s and 1790s.

Karl Friedrich Gauss made the breakthrough early in the nineteenth century while working on a method for the combination of measurements. From the "generally acknowledged" result that the arithmetic mean is the most probable value for a number of measurements of a single unknown, he showed that the errors of observation must follow a bell-shaped, or 'normal', distribution.[18] Gauss then applied the inverse argument to the general problem of fitting a curve of a given form to a series of data points. The best – in the sense of

---

[18] This distribution had been derived as an approximation to the unwieldy binomial expression by De Moivre in 1733, while working to reduce the limits of Jacob Bernoulli's law of large numbers. It was first tabulated by Christian Kramp in 1798, as part of his study of refraction effects in astronomical observations.

most probable – fit turned out to be that for which the sum of the squares of the differences between each data point and the corresponding value of the curve is a minimum.

Gauss's publication on this 'method of least squares' occasioned a disagreeable priority dispute. Adrien Marie Legendre had been the first to publish the least-squares coefficients, in a 1805 study of comet orbits, but Gauss claimed that he had invented the method in 1795, and had used it in 1801 to identify from scanty data the newly-observed asteroid Ceres. Certainly Gauss's discussion of the method, largely written in 1798 though not published until 1809, was more complete than that of Legendre. In particular, Gauss's work tied the normal distribution directly to error analysis and the method of least squares. This enabled him to present not only values for the least-squares coefficients, but probabilistic estimates for their errors too.

Despite its relatively complicated form – e^{-x^2} compared with the $e^{-|x|}$ distribution Laplace was using in 1774 – the assumption that errors were normally distributed actually simplified the mathematics of inverse probability. Laplace read Gauss's work around 1810 and was immediately provoked to rejoin the fray. First, he used his central limit theorem to generalize Gauss's justification of the normal law to any situation in which large numbers of small and independent causes operated. Then, in the monumental *Théorie Analytique des Probabilités* of 1812 – the most important single work in the history of probability theory – he presented a complete synthesis of the probability of causes and the analysis of errors.[19] The *Théorie* is innovative both mathematically and for its formal development of the inverse probability calculus as a method of rational judgment. Laplace showed how inverse probability and the normal law could be used in scientific matters to calculate the probability of various causes and their associated errors. As illustration, he assessed hypotheses concerning the form of orbit of a certain comet. He also showed that the method was especially helpful when applied to the weaker causes found in the social and moral realm, producing inverse probability distributions for birth-rates, voting procedures, and for all sorts of decision-making processes, including analyses of legal testimony and medical problems. "It is remarkable," he wrote, "that a science which began with the consideration of games of chance should have become the most important object of human knowledge ... The most important questions of life are, for the most part, really only problems of probability."

---

[19] The introduction to the *Théorie Analytique des Probabilités* was published separately as the *Essai Philosophique sur les Probabilitiés* in 1814.

Laplace's mathematical methods, and his clear association of a probability with degree of a belief, opened a wide new province for the doctrine of chances. In England, popularizers such as the insurance worker John Lubbock and, especially, the prolific mathematician Augustus De Morgan – who originated the expression 'inverse probability' – spread the word of Laplacean probability and its practical applications. In a tract, *On Probability*, co-authored with John Drinkwater-Bethune (though commonly attributed to De Morgan), Lubbock covered ground from horse racing to the probability of a given individual's telling the truth. De Morgan, who was in 1831 to resign his chair in mathematics at University College London because of the compulsory religious component of teaching, embraced inverse probability theory as a corrective to spiritual zeal. The recognition that different people formed judgments – probability assessments – relative to different bodies of background knowledge was the key to ending the "abomination called intolerance."

2.6. LAPLACEAN PROBABILITY

In 1814, Laplace defended the theory of probability as "good sense reduced to a calculus." But the status of his theory of probability depended on a particular view of man's relationship to knowledge. Like Pascal, Descartes, and indeed most of the philosophers and scientists of the time, Laplace regarded the material world as strictly deterministic. Jacob Bernoulli had predicted that exact calculations would one day make gambling on dice obsolete; Laplace declared more ambitiously that a sufficiently vast intelligence, possessing a complete knowledge of the state of the world at any instant, could predict with certainty its evolution through time. The Newtonian world-view was rampant in the early nineteenth century, and the universe commonly likened to a mechanical contrivance that once wound up by God would run according to strict and immutable laws with the reliability of clockwork.

Where, then, was the place for probability? Well, there was no need for such a concept – not at least for an omniscient God. Man, however, was an imperfect being, his senses too gross to detect the subtle causal mechanisms that underlay all material change. Though causes are indeed a matter of certainty, Laplace explained, our always imperfect knowledge of them is not. Like De Moivre and David Hume, who had dismissed the common belief in chance events as vulgar, Laplace regarded "chance as but an expression of man's ignorance." Probability is not in the world, but reflects our incomplete knowledge of the world. A judgment of probability expresses a mental state, a degree of knowledge or belief, and the calculus of probabilities governs the way that this knowledge is updated in the light of new evidence or experience.

The Laplacean view conflated the epistemic and frequency interpretations of probability. Though at bottom epistemic, a probability could when backed by the law of large numbers be measured as an objective long-run frequency. Laplace himself moved freely between interpretations, justifying least squares with both inverse and direct arguments, and applying the same calculus to testimony and scientific hypotheses as to dice games and insurance.

2.7. THE ECLIPSE OF LAPLACEAN PROBABILITY

The theory of errors developed rapidly in the first part of the nineteenth century. By around 1825, the method of least squares was almost universal among physicists. In part this was due to Gauss's renown as a mathematician, and the fact that in Germany, at least, many of his students became either prominent astronomers or carried the method with them to other fields of physics. It was also partially due to the growing professionalization of the discipline: a hundred years previously, such a rule-based treatment of data would have been dismissed as an impertinent encroachment on the expertise of the astronomer; during the nineteenth century it was welcomed as a corrective to the subjective aspect of experimentation, one that would inculcate the shared standards of professional conduct required for precise measurement.[20] With its new role in the calculations of inverse probability, the normal or Gaussian curve likewise soon became the dominant form of error distribution. Siméon–Denis Poisson's 1813 discussion of random variables implied that sets of data could be thought of as random samples from such a distribution, and by 1830 the normal curve – or by now, simply 'error law' – was widely

---

[20] Thus in 1895, Wilhelm Jordan frankly hailed the moral advantages of least squares: in preventing the fraudulent analysis of data, the method ensured honesty among geodesists and astronomers (Gigerenzer 1989, p. 84). This moral imperative worked in both directions. Objective statistical methods produced the 'exact sensibility' that arose in the physical sciences during the nineteenth century, but also required a tightly-controlled 'culture of experimentation,' in which individual creativity was effaced, and laboratory work reorganized into a series of regulated and routinized tasks to ensure that the conditions under which experimental data was generated conformed with those necessary for statistical analysis. Such forms of conduct needed to be standard if the precise numerical results obtained in one laboratory were to be accepted elsewhere; in the case of electromagnetic units, a common theoretical understanding of the effects under investigation was required too. For the relationship between precision measurements and experimental practices see, e.g., Swijtink 1987, Porter 1995, and the essays collected in Wise 1995, especially Schaffer 1995 and Olesko 1995.

assumed to describe not just errors themselves but the probability of individual measurements too.

Yet the success of the Gaussian error curve and the method of least squares concealed cracks in the Laplacean edifice. The inverse method as a solution to the problem of causes was not so secure. One immediate response to Laplace was to argue that elevated matters of philosophy and human knowledge could not be reduced to mere numbers and formulae. Many natural philosophers dismissed outright the balls-and-urn model as illegitimate for natural events. The rising of the sun was a different sort of thing entirely from the simple calculation of the probabilities in a well-defined game of chance.[21]

The assumption that probability provided a unique scale of rationality was also coming under fire. The classical probabilists had taken for granted that the calculus, as a description of the instincts and judgments of individuals of good sense, could be applied by plebeians as a prescriptive guide to rational thinking and behavior. Hence the stand-off over the St. Petersburg problem: Daniel Bernoulli argued that rational thought was exemplified by the preferences of shrewd dealers and could be captured by the concept of utility; the lawyer Nicholas Bernoulli, in contrast, insisted that reasonable behavior could only be underpinned by a concept of justice. From the beginning of the nineteenth century, however, and especially in France following the social upheaval of the Revolution, the underlying notion of a shared standard of reasonableness came to seem increasingly untenable, even ridiculous.[22] Although Poisson was still tinkering with the calculations to match intuition in 1837, the idea of probability as rationality was widely regarded by this time as a dangerous or foolhardy extension of the theory.

The reaction to the assumption of a common standard of rationality was evident also in new theories of cognition. David Hume, though wary of some predictive applications of probability, had, along with empiricist philosophers such as David Hartley and John Locke, promulgated a passive psychological theory in which countable sense impressions were physically imprinted onto the mind. But this associationist view, which supported the idea of a

---

[21] Neither Price nor Laplace was as naive as detractors presented. Both recognized that their calculations supposed a total ignorance of nature, and that a real observer would soon infer that the rising of the sun, as so many other natural phenomena, was not a probabilistic process but a consequence of some causal law. Price's point was to refute Hume's contention that some sorts of knowledge, though derived from observation, constituted "a superior kind of evidence . . . entirely free from doubt and uncertainty": even wholly uniform experience need not imply certainty, and seemingly law-like behavior could always be perverted. See L.J. Savage in Dale 1991; Zabell 1989a, pp. 294–6.

[22] Alder 1995 discusses the popular resistance in France to the imposition of the new 'rationalism,' in particular the resistance to the metric system of measurement.

single and objective measure of belief based on accumulated experience (and which blurred the distinction between epistemic probabilities and frequencies), had lost ground by the beginning of the nineteenth century, and instead psychologists were emphasizing perceptual distortion and idiosyncrasy.[23]

But there were also internal problems with the calculus arising from the strain of forcing a theory of causes and probabilistic inference into a model best suited for games of chance. Consider two events, e and h, which occur with probabilities $P(e)$ and $P(h)$, respectively. If these events are independent, the probability of both occurring together is simply the product $P(e)P(h)$. The probability of throwing a double-six with two dice is $\frac{1}{6} \cdot \frac{1}{6} = \frac{1}{36}$. In general, however, e and h might depend on each other. In this case the joint probability is the product of the conditional probability of e given h occurs, and the probability of h:

$$P(h \& e) = P(e \mid h)P(h)$$

where $P(e \mid h)$ denotes the probability of e given h. Of two children, the probability that both are boys, given that at least one is, is one-third, not one-half.[24] Now, from symmetry,

$$P(e \& h) = P(h \mid e)P(e)$$

and since $P(h \& e)$ clearly equals $P(e \& h)$, the expressions can be combined to obtain

$$P(e \mid h)P(h) = P(h \mid e)P(e)$$

or, rearranging,

$$P(h \mid e) = P(e \mid h)P(h)/P(e)$$

This is harmless as a relationship between the probabilities of events. But according to the classical probability of causes, it can also apply to hypotheses or propositions, conditional on background data. Let h stand for a hypothesis and e for a body of evidence. Then this core equation of inverse probability, now known as Bayes's theorem, expresses the probability of a hypothesis, h, given a body of data, e, in terms of the separate probabilities of the hypothesis and the evidence, and the probability of the evidence given the hypothesis. The conditional probability of h given e is to be interpreted as the degree of

[23] See Daston 1988, pp. 191–210. Daston quotes, p. 200, Hume in his *Treatise*: "Each new [experience] is as a new stroke of the pencil, which bestows an additional vivacity on the colors [of the idea] without either multiplying or enlarging the figure."

[24] See, for example, Grimmett and Stirzaker 1992, pp. 8–12.

rational belief in the hypothesis given the evidence. Thus Bayes's theorem governs the revision of knowledge in the light of new data.

Laplace termed the conditional probability of the hypothesis on the data, $P(h|e)$, the 'posterior' probability, and the probability of the hypothesis before consideration of the data, $P(h)$, the 'prior' probability. Bayes's theorem relates the posterior probability to the prior probability, but gives no indication of how this prior probability should be assigned in the first place. Given a series of observations, we can produce the factor that enhances the probability of the hypothesis under test, but to produce an absolute value of this new probability we need some indication of how likely our hypothesis was to start with.

What to use for these prior probabilities? Accustomed to the tradition in classical probability, in which problems were set up in terms of equally likely cases, Laplace tended to assume what became known as the Principle of Insufficient Reason – that where we have no knowledge of causes, or reason to favor one hypothesis over another, we should assign each the same probability. Thus in his study of comets, Laplace took all values of the orbit's perihelion – its nearest point to the Sun – to be initially equally likely. Likewise, his derivation of the Rule of Succession proceeded from the assumption that any number of black balls in the urn was initially as probable as any other.[25] In practice, Laplace would partition all possible outcomes into 'equally undecided about' groups, then with combinatorial analysis count those that led to some specified outcome, and finally take the ratio to the total number of cases.

The 'equally possible cases' of the classical definition were confusing enough in gambling conditions, where they had an intuitive grounding in the well-shuffled pack or unbiased die, and could readily be tested against abundant data. Jean le Rond d'Alembert wrongly thought that the three distinct outcomes in tossing a coin twice or until a head shows were each equally possible.[26] The classical definition was still more difficult to extend to the probability of causes. Well-specified causes can rarely be decomposed into 'equally possible cases.' Arguments from symmetry or ignorance are also rare. Indeed, the invocation of 'equally possible cases' seemed dangerously circular.

From the beginning of the nineteenth century, philosophers and mathematicians began to attack the Laplacean scheme. First was the argument

---

[25] De Morgan also made tacit use of the Principle of Insufficient Reason. Gauss also started from uniform prior probabilities in his derivation of least squares, but justified this assumption with an appeal to the equal distribution of empirical observations.

[26] He took the three cases H, TH, and TT to be equally likely, whereas the first is twice as likely as each of the second and third.

that ignorance cannot, by definition, ground knowledge. The philosopher John Stuart Mill in 1843 denounced the entire classical analysis as spurious: knowledge was based solely on experience; that Laplace had attempted to conjure conclusions relating to matters of fact from ignorance was nothing short of scandalous. "To be able to pronounce two events equally probable, it is not enough that we should know that one or the other must happen, and should have no ground for conjecturing which. Experience must have shown that the two events are of equally frequent occurrence."[27] The Reverend Richard Leslie Ellis repeated the charge the following year. *Ex nihilo, nihilo.* That the probability of causes had any credibility at all resulted from the spurious air of precision lent by quantification. Yet it was not clear, he pointed out, even how something like the Rule of Succession could be applied. Drawing from an urn is a properly-specified event, but what in real life counted as a repetition of a similar case? Even proponents of inverse probability shared this skepticism. In his book of 1854, *An Investigation of the Laws of Thought*, George Boole, like De Morgan, treated probability as a branch of logic, and thus applicable to the relationship between propositions.[28] But he denied that every proposition could be assigned a definite numerical probability with respect to a body of data.

As a consequence of these difficulties, the probability of causes started to lose ground while Laplace was still working. The first casualty was the testimony calculations. The French enthusiasm for selecting juries according to probabilistic considerations had if anything intensified in the years immediately following the invention of inverse probability. The new calculus was used to reason from the 'effect' of conviction rates and jury votes in split decisions to the 'cause' of a juror's reliability or the accused's probability of guilt or innocence, avoiding the need for at least some of the unrealistic assumptions of earlier work. Condorcet had again led the way, followed by Laplace and the more cautious Poisson. Yet attempts to apply these calculations in practice were quickly abandoned after the Revolution. The decisions of jurors could not be likened to the toss of a biased coin, despite Poisson's claims that his prior probabilities were based on measurements of objective chance. Moreover, prior probabilities could be assigned by the Principle of Insufficient Reason in a number of different ways, and this strongly affected the optimum jury size and voting procedure. Many probabilists selected priors

[27] Mill 1843, §2.

[28] Bernard Bolzano published in 1837 a precursor of the logical interpretation, in which probability, though reflecting a degree of belief, is primarily a relationship between propositions. Logical probabilities are objective in the sense that they depend on the propositions rather than any particular individual's consideration of them. See Dale 1991, pp. 286.

in accord with their prejudices or political sensibilities: though an aristocrat, Condorcet was a liberal reformer and was concerned to produce a system that would protect the innocent man wrongly accused, whereas the more conservative Laplace assigned his jurors prior probabilities that made it easier to convict the guilty. The whole exercise was largely discredited by the time Poisson published on the subject in 1837.

2.8. SOCIAL STATISTICS

The decline of the probability of causes is also a story of the independent rise of more objective approaches to probability. Ian Hacking has drawn attention to the "avalanche of printed numbers" that descended on Europe during the early years of the nineteenth century. As early as 1830, the word 'statistic,' which had been coined in Germany around 1750 to refer to verbal descriptions of the climate or geography of a region, had come to acquire an exclusively numerical sense. The act of counting or classifying is a way of imposing control, and following the efforts of amateur fact-gatherers, such as Sir John Sinclair and his massive *Statistical Account of Scotland* of 1799, and the private or friendly societies who used their numerical tables to press for reform in the classroom or workplace, it was chiefly the minor officials of new government agencies and bureaux who led the drive to classify and quantify. These men were the heirs of the old political arithmeticians. Just as a scientist needed to measure his system before he could understand it, so society in all its aspects needed to be reckoned before one could ask how it should be governed. Accurate data were essential for centralized administration, especially the organization of military service and the imposition and collection of taxes. The clerks of petty officialdom took to their work with zeal, supplanting the sporadic birth and death records of parish registers with accurate and regular censuses, measuring rates of crime, suicide, infection, and divorce, and compiling tables of trade figures and tax revenues. The urge to count was so great that categories and sub-categories began to multiply. The rapid specialization of census classification schemes is an example. Hacking notes that there were over 410 regular publications of government figures by 1860, when at the turn of the century there had been none.

It was not just the government agencies that embraced the wholesale quantification of society. The general public was also fascinated, in particular by the apparent stability exhibited by so many of the new indices. In a limited way, such regularities had been noticed earlier, but had generally been used as evidence of deterministic laws of nature, or to add force to the logic of natural theology. Thus John Arbuthnot in 1710 had argued that the stability of the male-to-female birth ratio was evidence of divine providence: the slight but persistent excess of male to female births was offset by male losses due to war so that exact parity was ensured for the holy institution of marriage. The German pastor Johann Peter Süssmilch similarly used population data in the 1740s to demonstrate God's order and to vindicate the ethical strictures of his Protestant faith. (Increased mortality in the cities was explained by the associated sinful opportunities for dissipation.) With age so easily quantifiable, there had been attempts to produce laws of mortality since the time of Graunt. But here, suddenly, were laws of murder, insanity, suicide, illness, and even absenteeism. They held with remarkable regularity from year to year and among groups by age, sex, and occupation. The existence of such regularities, generally unsuspected around the turn of the century, was a commonplace by 1834, when Henry Lytton Bulwer wrote that "Any one little given to the study of these subjects would hardly imagine that the method by which a person destroys himself is almost as accurately and invariably defined by his age as the seasons by the revolutions of the sun."[29] Even the mathematicians took note. Poisson used this 'fact of experience' in his legal calculations to license the transfer of an average value from empirical data to a numerical reliability for an individual juror.

These numerical laws of the early nineteenth century put a new emphasis on society. The actions of an individual might be too capricious or whimsical to be studied scientifically, but a collection of individuals was clearly a different matter. And a society was more than a simple aggregate of ideally reasonable men acting independently, but had a distinct character of its own. The explosion of numbers thus held the promise of a new way of learning about mankind. Students of the new sociology soon glossed the observed regularities as more normative than descriptive. Moreover, since the different causes that acted at the individual level seemed to result in gross stability, the assumption of a common or fixed set of social causes was not required. Poisson's justification was inverted, and his law of large numbers widely invoked as proof both of the apparent regularity of mass phenomena, and of the applicability of mathematical methods to study them. People had become statistical units.

The new social scientists embraced the statistical approach. Numerical data was trustworthy, since free of prejudice or political bias. In 1834, the first issue of the journal of the Statistical Society of London declared that the statistician was concerned solely with the facts, and not with opinions. The appearance of objectivity was at a special premium, given that many of the early statisticians

---

[29] Bulwer 1834, p. 203 as quoted in Hacking 1990, p. 74.

were attracted by the power of this emerging form of argument and persuasion precisely because they had some political or ideological agenda. Some appealed to the stability of statistical ratios to argue against government intervention: since society was an autonomous entity only partially sensitive to the actions of individuals, the prospects for ameliorative reform were bleak. Others took the opposite line. In France, especially, the statistical argument was often mobilized by liberals keen to wrest power from the church or royalty and place it instead in the hands of the state bureaucracy. Hence too the preponderance of data relating to social deviance of one form or other. That the act of counting is inherently dehumanizing in part explains the concentration on criminals, suicides, the insane – those regarded as barely human to start with. Counting can also be used as a means of persecution: Hacking records the role of statisticians in legitimizing anti-Semitism in Germany around 1880. But many of the data-gatherers focused on these unfortunates out of a sense of responsibility, intending to draw attention to their plight. Thus statistical tables could show the iniquities of alcohol or birth-control, or tie the incidence of disease to poor sanitation and hence prove the evil of slum housing.[30]

The most conspicuous proponent of the new scientific approach to society was the Belgian astronomer Adolphe Quetelet. His was a positivist program: the study of society was no different in essence from the lofty study of the heavens, and the aim of the social scientist was not to appeal to reductive mechanisms or underlying causes, but to extract from the available data of society laws that, like the Newtonian laws of gravitation, enshrined the observed regularities. Together with explicit scientific language and extensive physical analogies drawn from his astronomical background – he talked of conservation laws of moral force, for example – Quetelet imported the interpretation and mathematics of astronomical error theory. From around 1844, his 'social physics' began to rely on the Gaussian error law. Not only were ratios of birth or crime figures stable, but so too were entire distributions of continuous human characteristics. Of any reasonably-sized group of children, roughly the same proportion would fall into specified divisions of height or shoe size, or even of non-physical attributes such as mental ability or musical facility. Astoundingly, in each case the overall distribution seemed identical to Gauss's normal error curve. When his analysis showed the chest measurements of 5,738 soldiers of a Scottish regiment conformed to the Gaussian curve with remarkable accuracy, Quetelet became so convinced the distribution was universal that he declared a non-normal distribution of heights

---

[30] See Metz 1987.

among French conscripts proved that shirkers were falsifying their details to escape military service. Quetelet also reified the mean of the normal distribution. Where in the astronomical case the mean of the distribution represented the true value, and the scatter the errors of observation, so in the social case the mean must represent the human archetype, and the variation a sort of mistake due to perturbing factors. This interpretation carried a moral force. Quetelet, who after unpleasant experiences during the Belgian Revolution had become convinced of the value of moderation, likened his ideal individual – *l'homme moyen* – to the perfect form of the species, and declared that an actual population was made up of imperfect copies of this ideal.

From his introduction to the subject in the early 1820s, Quetelet worked tirelessly to champion quantitative studies and establish statistical societies. Despite the regularities in the data, he believed in slow but continuous social progress marked by the growing dominance of intellectual over physical ability. Such progress could be measured, since as aberration was reduced and individuals converged on the perfect form, the variation of any attribute in a population would narrow. In the meantime, that social distributions seemed fixed regardless of external forces need not imply fatalism. A robbery or murder, though carried out by a depraved criminal, was indeed a reflection of the entire society. Yet armed with Quetelet's tables, a government was free to select judiciously those policies that would shift the entire balance of society and thus reduce rates of deviancy.

2.9. THE RISE OF THE FREQUENCY INTERPRETATION OF PROBABILITY

The classical urn or gambling model of rational judgment implied a relationship between frequencies and degrees of belief. Since any rational man would form the same opinion when confronted with the same evidence, his subjective degree of belief could be identified with the objective aggregate or long-run frequency. The pressure on the doctrine of causes started to test this linkage, but with the advance of social statistics and the success of error theory in treating large quantities of numerical measurements, the strain became pronounced. Poisson in his *Recherches sur la Probabilité des Jugements* of 1837 distinguished between the two forms of probability, reserving that word for the epistemic sense, and using 'chance' to refer to the objective frequencies of events.[31] From then the epistemic interpretation lost ground rapidly. The

---

[31] In the same work, Poisson introduced his eponymous distribution, an approximation to the binomial for rare events. The distribution was later made famous by von Bortkiewicz,

growing emphasis on ratios in ensembles of data soon occasioned a reevaluation of the foundations of the theory. Within a few years in the early 1840s, a number of empiricists, including John Stuart Mill and Richard Leslie Ellis in England, Antoine Augustin Cournot in France, and Jacob Friedrich Fries in Germany, independently promoted frequency theories of probability. Each explicitly justified his approach with the evidence of the statistical regularities recorded by the social statisticians.

Grounding probability in frequency avoided many of the problems of the subjectivist approach. Probabilities, the frequentists insisted, applied only to events or measurements, not hypotheses or causes. Probabilities could be used for modeling and estimation, not inference or matters of judgment. Further, as a frequency or ratio, a probability referred collectively to the population, and could not be transferred to an individual item or member. (Thus Mill distrusted the practice of determining insurance premiums payable by individuals from demographic records.) The frequency theory also instituted a new statistical model of measurements and the process of measuring. Astronomical as well as social data were to be considered as random samples from a population of fixed distribution.

The original opponents of Bernoulli's art of conjecture had appealed to the argument that uncertainty could not be quantified or treated mathematically. With the rise of social statistics in the 1820s and 1830s, this argument was gradually displaced, especially among practicing scientists. Yet the frequency theory of probability brought a number of new and more powerful objections to the Laplacean view of probability as a degree of belief. The frequentists criticized not so much the formal development of Laplace's analysis as the assumptions on which it rested. Much of their odium was reserved for Laplace's Principle of Insufficient Reason, or, as the frequentists dubbed it the "equal distribution of ignorance."[32] Even conceding an epistemic role for probability, complete ignorance concerning two competing and exhaustive explanations could not be converted into the assessment that each has a probability of exactly one-half. If hypotheses could be assigned probabilities at all, it must be on the basis of frequencies in empirical data, not arbitrary mental

who compared it with the numbers of Prussian cavalry officers kicked to death by their horses. Though the chance of such an event is small, the large number of officers meant that one or more deaths per year per corps was not unusual, and von Bortkiewicz's data for fourteen corps between 1875 and 1894 fit the distribution closely.

[32] Several later authors, including R.A. Fisher, speculated that Bayes had been reluctant to publish because he recognized this problem, and that he had illustrated his formulation of inverse probability with the model of a ball rolling on a billiard table in an effort to sidestep it. Stigler, however, sees no hesitation in Bayes's exposition. Stigler 1986, 122–31.

states. Prior probabilities based on belief were at best subjective; they were simply indeterminate unless we had knowledge or belief to distribute. Boole brought up a second objection to the principle: it was inconsistent. Complete lack of knowledge about some parameter implies also ignorance about any function of that parameter. Yet a uniform distribution in one case will lead to a different probability distribution in the other. Knowing nothing about the density of an object implies ignorance also of the specific volume. Yet these quantities are reciprocally related, and cannot both be uniformly distributed. Continental thinkers were no less critical. The economist A.A. Cournot wrote that probabilities had little value other than the "charm of speculation" unless based on experience; Fries agreed.

2.10. OPPOSITION TO SOCIAL STATISTICS
AND PROBABILISTIC METHODS

The frequency theory of probability drew much of its support through the nineteenth century from the rise of statistical thinking. But this rise did not go unopposed. Medical applications of statistics were particularly controversial. Daniel Bernoulli had used statistical techniques in 1760 to demonstrate that despite the small but undeniable risks attending the method, smallpox inoculation increased overall life expectancy; Voltaire had concurred with this assessment, and the matter had been settled by Jenner with his cowpox vaccination in 1798.[33] But few other medical conditions were as susceptible to a clear statistical demonstration. The growing popularity of pathological examination in France following the Revolution afforded opportunities for objective measurement, and physicians such as Pierre Louis published from the 1820s a series of statistical inferences concerning various diseases and their suggested therapies. Yet as late as 1835 the opinion of the French Academy of Sciences was that the hard, numerical facts of medicine were too sparse to allow a valid probabilistic or statistical assessment.[34]

Such opinions were not a call for more data. On the contrary, notable physicians such as Risueño d'Amador and François Double were typical of their

[33] D'Alembert, though generally in favor of inoculation, criticized Bernoulli's calculations as genuine guides to action on the grounds that notwithstanding the lower mortality rates of the inoculated, it was not necessarily unreasonable for an individual to prefer the greater but long-term risk of contracting smallpox over the smaller but more immediate risks associated with inoculation. See Daston 1979.

[34] One of Louis's studies questioned the efficacy of blood-letting as a cure for cholera. Though the method, and its tireless advocate, the leech doctor F.J.V. Broussais, were indeed soon discredited, Hacking suggests this was due to factors other than the statistical arguments.

profession in resisting the efforts of medical quantifiers like Louis. These medics opposed the encroachment of statistics, and indeed quantification more generally, for reasons similar to those of the gentlemen insurers of the previous century. Medicine was a matter of fine judgment and a trained intuition, not equations and rule books. The numerical and statistical approach would not bring objectivity or rigor, as claimed its proponents, but would obscure the subtle effects revealed by expert scrutiny. Moreover, although by the mid-nineteenth century the frequency interpretation of probability was dominant in the social sciences, its implications appeared particularly insidious in the area of medicine. What cares the bed-ridden pox victim that in only ten percent of cases is the disease fatal? His own condition is unique. He is not interested in the average response to therapy across age group and social class, but purely in his own chances of survival. The statistical method requires gross comparability. But though it is possible to group, say, all coal-miners together for the purpose of calculating an insurance premium, each medical case is substantively different. Double appealed to the teaching of Hippocrates. Treatment was not a matter of 'one size fits all,' but should be gauged anew following painstaking observations of each patient's condition and reaction to treatment.

Positivist-minded physicians such as Claude Bernard and Carl August Wunderlich disagreed with this assessment. Yet they too were unimpressed by the potential of statistics. Medicine is indeed a matter of hard empirical evidence, they argued, but statistics provide merely an approximate description. At best, statistics can serve to pinpoint areas that might repay a more minute examination when underlying causes are unknown. Physicians would be better advised, however, spending their time in the laboratory studying physiological processes and uncovering genuine – that is, determinate and causal – laws of medical science.[35]

Nor were statistical methods wholly welcome in the developing field of sociology. Though Quetelet's influential social statistics were popularly believed to be a successful extension of the scientific method to the human and even moral spheres, some sociologists, including the founder of positivism, Auguste Comte, deplored the reductionism inherent in quantification. Mankind could not be expressed simply as number; instead, the richness of

---

[35] The argument was protracted and heated on all sides. Statistical methods, where they appeared at all, were primarily mobilized as polemical weapons – used "more out of professional jealousy than in a quest for objective knowledge," says Hacking 1990, p. 81 – and did not become an uncontroversial part of the medical discipline until the introduction of clinical trials in the early twentieth century.

human interaction must be understood through detailed historical study. Further, the statisticians' concentration on mean quantities obscured the process of social evolution, which was most heavily influenced by those of exceptional wealth, talent, or power, and which involved, as far as Comte was concerned, transitions between essentially incommensurable stages of social development. Numerical averages lumped together this fine structure, and elided the variation and differentiation necessary in a functional society.[36] Besides, added the French economist J.B. Say, statistics is a matter of description only, whereas a mature sociology should surely be aiming to predict social change.[37]

Popular commentators, particularly in England and Germany, were also growing uneasy with the non-mechanistic form of causation implicit in Bernoulli's balls-and-urn model. Despite his unequivocal statements of determinism, Laplace had not doubted the doctrine of free will; as Kant had written, thoughts were insubstantial and thus not bound by material necessity. But if society were really subject to a law that prescribed a certain number of, say, suicides each year, how could each individual be fully free in his actions? As well as insisting that statistics such as the murder and suicide rates were properly characteristics of a population rather than an individual, Quetelet tried to distance himself from any categorical statement of statistical determinism. The law of large numbers showed that the effects of free will, though important for an individual, would average out at the macro level, he counseled; sociologists should concern themselves with aggregate behavior only. The chess-playing historian Henry Buckle, however, had few such reservations. His immensely popular *History of Civilization* of 1857, hailed as demonstrating the extension of the scientific method into domains thought previously unpredictable, also became notorious for its shameless statistical fatalism. The natural order of society progressed according to iron laws inbuilt; free will was an illusion.

This debate continued through the 1860s. Yet even the most enthusiastic social quantifiers were starting to recognize that not all their data could be represented as even approximately normal. Ellis had already sounded a

---

[36] Comte had developed the idea of a hierarchy of knowledge, and thus also took issue with social statistics on the grounds that the methods and techniques proper to a study of society should be particular to it and not borrowed wholesale from the 'higher' disciplines of astronomy and physics.

[37] As in the medical case, the concept of social statistics was sufficiently plastic to provide plenty of rhetorical ammunition for both sides of the debate. On the one hand, a statistical or probabilistic theory could confer the credibility of mathematical rigor; on the other, such a study dealt in uncertainty and so by nature was vague.

dissenting note in pointing out that the assumptions necessary for a mathematical derivation of the normal law militated against its universal application in the human realm. It subsequently became clear that Quetelet's 'laws' varied alarmingly and were highly sensitive to particular conditions of place and time. Moreover, recidivism rates indicated that crime, far from being a characteristic of the entire society, was largely the work of specific individuals. Starting in the 1870s, the German economist and statistician Wilhelm Lexis developed a number of complex mathematical tests, including his 'index of dispersion,' to show that the atomic, balls-and-urn model of independent individuals operating subject to fixed laws of probability compared poorly with the fluctuations observed in real data. Apart from the sex ratio, social indicators such as the rates of suicide or crime were not particularly stable from year to year. Nor should their even approximate constancy be considered remarkable. Statistics, he argued, though a vital tool for the social scientist, should best be restricted to properly independent ensembles of data.[38]

2.11. PROBABILITY THEORY IN THE SCIENCES: EVOLUTION AND BIOMETRICS

The discipline of statistics developed in the nineteenth century along with a particular view of the place of the individual in society and the causes of social change. When, toward the end of the century, statistical methods started to be transferred to the substance of the physical and biological sciences – as opposed to just the analysis and reduction of measurements – they carried a different sort of implication. This shift has been well essayed by several recent writers, particularly Daston, Hacking, and Porter. In the sphere of human action, seen traditionally as chaotic or at least capricious, the mathematical theory tended to be welcomed as granting a degree of control and predictability. This is what Hacking means by his phrase "the taming of chance." But in the harder sciences, the use of probabilistic methods was resisted as introducing the taint of uncertainty into a branch of knowledge previously regarded as determinate and secure.

The interpretation of probability in the physical sciences during this period is considered in more detail in chapter six. In short, though, the precise significance of the theory in any particular case was complex. James Clerk Maxwell

---

[38] Lexis was responding to the determinism of Buckle. Lexis believed Buckle's historical 'laws' to be spurious, since if a causal force operates at the level of society it must to some degree be felt by the individual too. Yet there is no impediment to free will, Lexis argued, or to the resultant change and organic growth of society.

began his work on statistical physics in 1859 to see whether Quetelet's normal law could be transferred from the large numbers of people in a society to molecules in a gas. By the early 1870s he was using this 'kinetic theory' to argue that since knowledge of the physical world was always statistical in character, it could never be complete. Maxwell was a deeply religious man, even given the standards of the day, and he took this argument to allow room for the action of free will without violation of natural law.[39] Ludwig Boltzmann, on the other hand, was a strict determinist. He embarked on his study of statistical physics in an attempt to reconcile his mechanistic philosophy with the irreversibility of the second law of thermodynamics, which expresses the observation that heat always flows from hot to cold bodies, and never the other way. For Boltzmann, the use of probability theory did not imply uncertainty but was merely a convenient way of describing a large number of independent molecules.[40]

Since the connection between probability and incomplete knowledge is less tangible for living than physical systems, the situation was still more confused in the biological sciences. Thus on the one hand, any appeal to chance was rejected by many psychologists, who saw it as a cloak for vitalism; on the other, it was welcomed by naturalists as an antidote to teleology and arguments from design. Such commitments were also entangled with the battles of Victorian scientists at the end of the century to establish the autonomy of their profession and the force of scientific arguments independent of dubious foundations in natural theology. Many of the most vocal scientists in this struggle, such as John Tyndall and T.H. Huxley, were uncompromising in their deterministic mechanism. They recognized the danger of probabilistic arguments, such as Maxwell's, which allowed room for free will and thus could be mobilized to leave science forever subservient to the authority of the church.

In this context, it is understandable that the rapid expansion of statistical techniques and methods around the turn of the century is chiefly associated with the study of evolution. Darwin based his theory on chance variation, but took this to mean that causes of variation were unknown rather than random, and therefore did little to test it within a statistical framework. Such a research program was instituted by his cousin, the self-taught polymath

[39] Maxwell, who greatly admired Quetelet's work, was unusual in regarding the regularities of social measurements as evidence of free will. Free men, he argued, tended to act in a uniform manner. It was instead constraint or "inference with liberty" that produced disorder.

[40] In order to calculate probability ratios with combinatorial analysis, Boltzmann made the assumption that the energy distribution of his atoms was not continuous, but came in discrete energy levels. This later became the basis of the quantum theory.

Francis Galton, to whom Darwin passed data on cross- and self-fertilization in plants for analysis. Galton had come into a sizable inheritance shortly after graduating in medicine at Cambridge. Following a period of exploration in Africa, which resulted in a couple of popular guides to adventurous travel,[41] and recognition by the Royal Geographical Society, he embarked on a life of gentlemanly science. In 1858, he was appointed director of Kew observatory, where he studied weather maps and meteorological records. The exposure to quantities of numerical data rekindled an earlier interest in statistics. Galton turned his attention from meteorology to biology following the publication in 1859 of Darwin's *Origin of Species*. He focused in particular on physical and mental traits in humans.

After his introduction to Quetelet's work in the early 1860s by a geographer friend, William Spottiswoode, Galton started to apply the normal distribution to a wide range of observations. Though sharing, and perhaps even surpassing, Quetelet's almost pathological enthusiasm for this curve – the economist and statistician Francis Ysidro Edgeworth named this disease 'Quetelismus' – Galton differed with his interpretation. Quetelet had largely ignored variation, seeing it in astronomical terms as a sort of error and focusing instead on the mean of the distribution as representing the ideal type. For Galton, however, variation was not a trivial matter to be eliminated. In biological systems, not only was variation a real and measurable feature, it was the main motor of adaptive change.[42] The average was mediocre; it was the few exceptional individuals in a population who drove evolution.[43]

Galton was interested more in heritable effects and the spread of variation – especially in human populations – than in the hereditary process itself, and adopted pangenesis as a model of inheritance. Introduced as a provisional theory by Darwin in 1868, pangenesis explained reproduction in terms of 'gemmules,' which were produced in vast numbers in each part of the body and combined from both parents to form the embryo. Galton had little time for the implications of the theory – which seemed to support Lamarck's thesis that characteristics acquired during an organism's lifetime could be passed on to its progeny – but saw that the transmission mechanism was ideal for a statistical treatment. Offspring were formed from a blend of gemmules

---

[41] Including such advice as: when caught in a sudden downpour, remove all clothing, fold into a tight bundle, and sit on it until the rains have passed.

[42] Maxwell's introduction of the velocity distribution in kinetic theory also helped to shift the focus to variation, significant as an objective and measurable feature of a gas rather than merely as, say, a reducible artifact of errors in observation and instrumentation.

[43] Compare with Nietzsche, who in 1885 wrote in *Beyond Good and Evil*, §126, that "a people is a detour of nature to get six or seven Great Men."

selected from each parent like balls drawn from an urn. Using this model, Galton developed many new statistical techniques expressly for problems in heredity. Differing from Darwin in believing species to be distinct and essential, he introduced regression analysis in 1876 to quantify the reversion, or 'regression,' towards the ancestral mean of a trait in an offspring generation. Believing that laws of inheritance would be universal, he had studied the weights of successive generations of peas grown in batches for him by friends. He found that the width of the normal distribution of weights in an offspring batch did not depend on the parent peas, but that the mean weight lay between the overall mean and the mean of the parents. The extent of this reversion toward the ancestral mean, Galton discovered, was proportional to the parental mean. Galton used 'deviation' as a measure of a individual's aberration from the species archetype. He came to a statistical formulation of the idea of 'correlation' when applying similar ideas to human measurements in 1888. The qualitative fact that some traits are interdependent – men with big feet are likely to be tall, for example – had been used by Cuvier in his reconstruction of skeletons from only one or two bones, and could be explained within pangenesis by the idea of 'mutual affinities' between gemmules. Yet the French criminologist Alphonse Bertillon recommended identifying criminals according to the sizes of their body parts, assuming that finger, foot, and arm measurements were independent. Provoked by the redundancy in Bertillon's method, Galton found that by scaling the measure of each trait by the probable error he could produce a clear mathematical index of the correlation effect.

Galton instituted the 'biometric' school. Its aim was to use quantitative data of inheritance and variation, gathered from large biological surveys, to demonstrate clearly Darwin's controversial theory of natural selection. But this work also had a social agenda. Galton used genealogical trees, together with a series of creative analogies drawn directly from social physics, to argue that exceptional physical and intellectual ability in humans – genius, most notably – were hereditary traits rather than a product of environment. His own tree, including Charles Darwin and the Wedgewood family, he regarded as a case in point.[44] Yet his data also showed that this prize stock was being diluted. The advent of civilization had brought a suspension of natural selection in human society, Galton concluded, leaving on the one hand an aristocracy that was degenerate and inbred but that controlled wealth and political power, and on the other an inferior lower and criminal class of frightening fecundity.

[44] See Galton 1889. Edgeworth was a distant relative, as, subsequently, was the composer Ralph Vaughan Williams.

A statistical theory of inheritance was to be part of the call for social reform through selective breeding.

The biometric program was taken up by Karl Pearson.[45] Although graduating from King's College, Cambridge in 1879 as third Wrangler in mathematics – that is, third of those awarded a first-class pass in the Mathematical Tripos – Pearson had intended to follow his barrister father into the law. However, these plans were abandoned shortly after a period spent in Germany studying a range of progressive texts. Pearson had already demonstrated free-thinking tendencies: influenced as an undergraduate by Spinoza's philosophy, he had renounced established Christianity and campaigned against compulsory divinity lectures. But it was during his time in Germany that he began to condense his thinking in politics, theology, and history into a single perspective. Adopting the German spelling for his first name – previously he had been Carl – he returned to England and started to publish and lecture widely in these and other areas. In 1883, he divided his time between a temporary position in mathematics at King's College London and lecturing at 'revolutionary clubs' in Soho; the following year, and largely on the strength of his radical connections, he was appointed to a chair in mathematics at University College London.

Pearson held that Victorian morality stood in the way of the social good. Sexual mores, and in particular the established church, were stifling the welfare of society. Pearson's politics were strongly influenced by the writings of Karl Marx, and in some ways were similar to those of contemporary Fabians. Yet they would be barely recognized as socialist today. His "social good" derived primarily from the elite class of thinkers and professionals, and the rigidity of the class system was to be deplored not because it oppressed the workers, but because it obstructed the advancement of middle-class intellectuals such as himself. Pearson called for a sort of state socialism, in which a powerful government would control capital and welfare for the general good.

Evolution, not revolution, was Pearson's preferred route to social change: he was as anxious to prevent the uprising of the brute proletariat as any peer or churchman. When introduced to Darwin's theory of evolution in the late 1880s, he quickly saw that it provided a perfect naturalistic framework for his views on social history. But Darwin's theory did not just explain society, it gave a dire warning for the future of humanity. Pearson first encountered Galton's work through his *Natural Inheritance* of 1889, and was lecturing on the subject

---

[45] For Pearson see Yule and Filon, 1936; Norton 1978; MacKenzie 1981, chapter 4; Porter 1986, pp. 296–314; Pearson 1911; Pearson 1920b; Haldane 1957; E.S. Pearson 1965.

later that same year.[46] Traditional middle-class values, which discouraged breeding, must be swept aside. And clearly, education and improved living conditions for the poor were merely temporary amelioration that could not improve the worth of subsequent generations. Pearson's powerful central government was to preserve the potent and vigorous stock of the middle classes through a program of selective breeding, encouraging not marriage but eugenic couplings.[47]

In 1891, the University College zoologist Walter Frank Raphael Weldon approached Pearson for help in the analysis of biological data, and convinced him that "evolution was a statistical problem." Pearson delivered a series of popular lectures on the subject of probability and statistics from 1891 to 1894 at Gresham College, and started teaching the theory at University College in 1894. He also began to develop his own statistical techniques. Though initially sharing Galton's reverence for the normal curve as representing stability, Pearson saw that other forms would also arise in nature, and would have important evolutionary consequences. For example, Weldon's measurements of crabs in Plymouth Sound, gathered laboriously over a three-year period, gave rise to a distribution with two humps. Pearson's 'method of moments' of 1895 used a family of skew distributions to approximate and thus summarize the natural variation in such cases. He developed the more famous χ^2 goodness-of-fit test in 1900 to gauge how well skew multinomial distributions fit real data.[48]

Following Ernst Mach in Germany, Pearson promoted an extreme form of positivism. In his *The Grammar of Science* of 1892, based on the lectures at Gresham College, he argued that objective reality was a meaningless concept, and external entities nothing more than products of the mind, constructed to coordinate and order the data of sensations. ("No physicist ever saw or felt an individual atom.") Talk of cause and effect is a fetish, he declared, used simply to dress up observed regularities. Science *explains* nothing; it is merely the method by which experiences are classified and relations between

---

[46] Like Pearson, Galton had little respect for the church. In a notorious study of 1872 he argued that prayer was inefficacious, since priests and sovereigns – presumably the object of much prayer – lived on average no longer than lawyers or doctors. See Forrest 1974.

[47] For eugenics, see Kevles 1985. For the relationship between eugenics and statistical practice, see MacKenzie 1981.

[48] Influenced by Edgeworth, Pearson came to see asymmetric distributions as more common in natural variation, and hence more useful, than the normal curve. He worked extensively on the properties of a family of asymmetric distributions, and as Porter notes, came to evince "a faith in the reality of his skew curves that belied his militant positivism." Porter 1986, p. 309.

them discovered.[49] Galton's biometrics exemplified this positivism. A well-ordered summary of the facts and the correlations between them, rather than a speculation on causes, was required to show that Darwinian natural selection could account for evolution, and thus highlight the need for eugenic policy. Pearson regarded his political endeavors as sharing the same philosophy. He declared that the "unity of science is its method" – thus justifying the free movement of his own studies across academic boundaries – and regarded the theory of inverse probability, as a general way of combining evidence, as constituting this method. Statistics – not religion or philosophy – was the sole route to reliable knowledge. Scientists were to be the high priests of Pearson's new order. The scientific method would guarantee a uniformity of judgment and thus social cohesion. It was the duty of the citizen to master this method. Only then, when cleansed of prejudice and emotion, could he properly judge the duties of the individual and responsibility of the state.

2.12. THE INTERPRETATION OF PROBABILITY AROUND THE END OF THE NINETEENTH CENTURY

The rise of statistics during the nineteenth century was not accompanied by similar advances for pure probability theory. From the 1820s, error theorists such as Frederick Bessel had started to calculate parameter confidence limits; subsequently, actuaries such as the German Theodor Wittstein, and the academic economists Lexis, Edgeworth, and, to an extent, William Stanley Jevons, introduced a degree of mathematical rigor to the analysis of social data. (Both Wittstein and Lexis studied statistical fluctuations and the stability of statistical series, while Edgeworth developed early forms of significance tests, and in 1885 calculated the probabilities of various sorts of psychic effect.) Yet Joseph Fourier, who compiled statistical tables for the French government in the first half of the century, was unusual among social statisticians in presenting alongside his data probabilistic estimates of error, taken from a comparison with the normal curve. The majority of statisticians and biometricians regarded the use of mathematical methods to move beyond the bare data of numerical surveys to probabilistic inferences about entire populations as contrary to the original statistical spirit of presenting hard facts not

---

[49] The once-influential *Grammar* was reissued in 1900, and a third and final edition appeared in 1911. It prompted Mach to dedicate his *Science of Mechanics* to Pearson. V.I. Lenin denounced Pearson's politics as muddled, but thought the *Grammar* a powerful response to materialism. (Haldane 1957.) The title of the book came from Pearson's quarrel not with the results of science but with its language.

opinion. Hence government statisticians pressed for complete censuses to be taken, and regarded sampling as mere speculation.[50]

The neglect of probability theory also blurred the relative strength of the frequentist and epistemic interpretations. As I have described, the advance of statistical thinking did not go completely unchecked. Yet statistics, and its attendant frequentism, was generally successful through the nineteenth century as a means of getting a purchase on the apparent unpredictability of human behavior. It gave authority to social quantifiers, and resulted in a new form of argument – and indeed, culture of order and reliability – in which objective numbers became the final arbiter. The intense discussion of Buckle's book, which was cited widely from John Stuart Mill to Karl Marx, serves as illustration. From the 1850s, legislators and policy-makers paid increasing attention to statistical arguments, and by the time Durkheim published his celebrated study of suicide in 1899, many of his conclusions were truisms. The frequency interpretation of probability, furthered also by the work of Maxwell and Boltzmann in kinetic theory, was cemented in John Venn's influential *The Logic of Chance* of 1866. A respected mathematician and logician, Venn insisted that probabilities must only apply to large ensembles, since a single event could fall into several different categories. (What is the life expectancy of a tubercular authoress from Edinburgh, given that the groups of women, Scots, authors, and those with tuberculosis live on average to different ages?) Probabilities were properly frequencies in such ensembles, and could not therefore justify induction from single cases; still less did they apply to belief in hypotheses or causes. Dr Venn devoted an entire chapter of his book to ridiculing conclusions drawn from the Principle of Insufficient Reason, in particular Laplace's Rule of Succession. As he wrote, "The subjective side of Probability, therefore, though very interesting and well deserving of examination, seems a mere appendage of the objective, and affords in itself no safe ground for a science of inference."[51]

Yet the old epistemic view was not completely abandoned. Though the probability of causes started to recede from the literature around 1840, the debate over the correct interpretation and use of probability continued through

---

[50] Porter has noted in this regard that most practicing statisticians, whether in government offices or private societies, were insufficiently skilled in mathematics to originate or even use probabilistic theory. It was only with the efforts of the Russian school, starting with Victor Bunyakovsky mid-century and continuing with Pafnutii Tchebychev and A.A. Markov, that probability theory was eventually shorn of its social and astronomical applications and established at the end of the nineteenth century as a distinct part of pure mathematics.

[51] Venn 1962 [1866], p. 138. On the continent, Bertrand's 1889 attack on inverse probability was highly influential.

Victorian times. The purple rhetoric and willful misrepresentation on both sides indicate that the matter penetrated to core beliefs about the nature of knowledge.[52] Despite the objections raised from Mill to Venn, and the success of statistical thinking and the kinetic theory, many philosophers and logicians continued to regard chance as a reflection of imperfect knowledge. William Donkin published an astute defense of degree-of-belief probability in 1851, showing that with the Principle of Insufficient Reason as an axiom, the equation of inverse probability gave a satisfactory fit with several observed features of the scientific method. For example, under Bayes's theorem a theory is made gradually more probable with successive verifications, and is completely invalidated by a single falsification.

The theorem also gives the intuitively pleasing result that unexpected consequences of a hypothesis, if observed, lend it more weight than commonplace consequences. Hence too the importance of crucial tests in the history of science: an observation predicted by only one of two competing hypotheses will greatly raise the relative probability of that hypothesis, even if initially less probable. Even the influence of prior probabilities, deplored by frequentists because introducing subjective judgment, seems to accord with commonsense and practice: as evidence accumulates, posterior probabilities of a hypothesis will converge whatever might be their prior assessments.[53]

Many of the defenses of inverse probability were nuanced. George Boole, who extended Donkin's work in his 1854 book and subsequent writings, deplored naive applications of the Principle of Insufficient Reason. Jevons also preferred to think of probability as a relation in logic, reflecting not the subjective-sounding 'degree of belief,' but a 'quantity of knowledge' that was always provisional. Yet he explicitly likened science to observing the draws from the infinite ballot-box of nature, and used the Principle of Insufficient Reason freely. He illustrated how inverse probability could account for the process of induction, and thus the scientific method, with examples from astronomy in his 1874 book *The Principles of Science*. In contrast, the indefatigable De Morgan, popularizer of Laplace's probability analysis in England, rejected the extension of probability theory from the repetition of events, 'pure induction,' to the induction of general laws of nature. But he continued to champion inverse probability up to his death in 1871.

---

[52] Zabell, pointing out that opponents seemed more concerned to ridicule than persuade each other, has showed how the debate over the Rule of Succession revealed preconceived notions of quantification and the role of probability in knowledge. Zabell 1989b.

[53] Appendix 2 is a fuller account of inverse probability as a model of the scientific method.

Even if few attempted to apply Bayes's theorem in other than a qualitative way, many practicing scientists also took the calculus of inverse probability to govern the process by which theories or hypotheses are tested in the light of new evidence. Thus in 1849, the president of the Royal Astronomical Society, John Herschel, assumed initial equiprobability in his textbook *Outlines of Astronomy* to assess the probability that closely-spaced stars were evidence of binary systems, rather than simply consequences of random scatter. Despite objections to this work,[54] Herschel's defense of inverse probability for real scientific predictions was so authoritative that Mill diluted his criticism of the Principle of Insufficient Reason for the second edition of his *A System of Logic* of 1846. The same was true even of some statisticians. As a committed positivist, Karl Pearson eschewed the language of causes, instead regarding statistical description as the sole kind of knowledge. Yet he interpreted a probability to represent a numerical degree of credibility, and regarded inverse probability as the proper account of learning from experience. And although Pearson followed Edgeworth in propounding a strictly 'empirical' form of Bayesianism – that is, insisting that prior probabilities should only be assigned from the experience of statistical ratios – he occasionally resorted to the Principle of Insufficient Reason to justify an initially uniform distribution. Examples are a 1898 paper co-authored with L.N.G. Filon on the probable errors of frequency constants, and a general 1907 paper on inference from past experience.[55] Gosset too defended inverse probability, occasionally guessing at forms of prior that could reasonably represent experience. His use of polynomial priors in 1908 is an example. Maxwell embraced epistemic probability with enthusiasm as the model of inductive reasoning, writing that "the true logic for this world is the calculus of Probabilities, which takes account of the magnitude of the probability which is, or ought to be, in a reasonable man's mind." The physicist Henry W. Watson originally dismissed Galton's formulation of correlation on the grounds that it violated the laws of inverse probability.[56] Though of course rejecting both the epistemic and logical interpretations of probability, even Venn occasionally wavered. Considering a small-sample example concerning Lister's antiseptic system, he conceded that there was often no alternative

---

[54] For example, by Boole and J.D. Forbes. See Dale 1991, pp. 71–8; Porter 1986, pp. 118–24.
[55] See Edgeworth 1910. Boole too regarded the calculus as safest when an equal distribution of probability was justified from empirical data. Then the "principle of cogent reason" contrasted with the "equal distribution of ignorance." Johannes von Kries likewise defended basing the distribution on statistical data or from arguments of symmetry. See Hacking 1990, Porter 1986, Dale 1991, Stigler 1986, Zabell 1989a, 1989b.
[56] See Porter 1986, pp. 294–6.

to the Principle of Insufficient Reason as a starting point in many practical situations.[57]

Epistemic probability survived in financial circles too. Though the mathematician George Chrystal advocated the frequency interpretation to the Actuarial Society of Edinburgh in 1891,[58] insurance men such as George Hardy, John Govan, and William Makeham continued to calculate mortality probabilities with the Rule of Succession in the 1880s and 1890s. (Indeed, despite the usual objections to the formulation of prior knowledge, E.T. Whittaker was successfully defending inverse probability to actuaries in 1920.[59])

The general position is illustrated by the article on probability in the ninth edition of the *Encyclopædia Britannica* of 1885. The author, Morgan Crofton, starts with the statement that the "probability, or amount of conviction accorded to any fact or statement, is thus essentially subjective, and varies with the degree of knowledge of the mind to which the fact is presented," before explaining that "the knowledge we have of such probability in any case is entirely derived from this principle, viz., that the proportion which holds in a large number of trials will be found to hold in the total number, even when this may be infinite."[60] For the majority of scientists and philosophers, then, the objective and subjective interpretations of probability remained entangled into the twentieth century. Few scientists attempted to evaluate their hypotheses with the probability calculus; when they needed to analyze experimental data, they turned to the methods of statistics, and the attendant frequency interpretation of probability. Even committed Bayesians recommended assigning prior probabilities according to statistical ratios. Yet most scientists acknowledged the epistemic interpretation of inverse probability as a model of scientific inference. The French mathematician Henri Poincaré was clear: "all the sciences are only unconscious applications of the calculus of probabilities. And if this calculus be condemned, then the whole of the sciences must also be condemned."[61]

---

[57] The antisepsis problem was originally raised by the Cambridge mathematician Donald MacAlister in the *Edinburgh Times*, and provoked much discussion. See Dale 1991, pp. 338–42.

[58] Chrystal was "merely one of many ... whose intellectual attainments in other areas led him to uncritically accept his own untutored probabilistic intuitions," remarks Zabell 1989b, p. 252, apropos Chrystal's misunderstanding of Bayes's theorem.

[59] See Dale 1991, pp. 366–9.

[60] Quoted in Dale 1991, pp. 348–9.

[61] Poincaré 1952 [1905], p. 186.

3

R.A. Fisher and Statistical Probability

Ronald Aylmer Fisher was born in 1890 in East Finchley, London. An academic prodigy at Harrow, his interests ranged from mathematics and astronomy to biology and evolution. His mother died while he was still at school, and shortly afterwards his father, an auctioneer of fine art, bankrupted himself through a series of disastrous business deals. The once-prosperous family fell on hard times. In 1908, however, Fisher won a scholarship to Cambridge. Though the rote nature of the degree course in biology led him to plump for mathematics instead, the young Fisher continued to read widely, and finishing the three years of mathematics, returned to Cambridge in 1912 on a one-year physics scholarship.

In this final year, Fisher attended lectures by James Jeans on the quantum theory, and by his college tutor, the astronomer F.J.M. Stratton, on the theory of errors. But by this stage he had discovered a perfect combination of his mathematical and biological interests in Karl Pearson's "mathematical contributions to the theory of evolution."[1] Fisher was especially attracted to Pearson's focus on human hereditary, and took to eugenics with idealistic zeal. With a group of like-minded friends, he was instrumental in establishing a Cambridge University Eugenics Society – John Maynard Keynes was another founding member – and led discussions of Pearson's ideas and their implications for society. Fisher acted as a steward at the First International Eugenics Conference in 1912, and addressed the larger London Society in October of the following year.[2] In his reading of history, the decline of civilizations was linked to a differential rate of birth between the classes. Like Pearson, Fisher advocated 'positive' eugenics, with an emphasis not so much on stemming

---

[1] Pearson used this one title for many of his memoirs on the subject, having used it in a slightly different form – Contributions to the Mathematical Theory of Evolution – for his first statistical paper of 1894.

[2] Fisher 1914.

the inherited causes of feeblemindedness and criminality – though these were concerns – but in recognizing and encouraging the superior genetic stock of the middle and professional classes. But unlike Pearson, Fisher was as interested in the biological causes of inheritance as the effects.

3.2. EVOLUTION – THE BIOMETRICIANS VERSUS THE MENDELIANS

Though the fact of evolution was widely accepted in the scientific community – and had been for many years before the publication in 1859 of Darwin's *Origin of Species* – the causal mechanism was still a matter of contention. Darwin had focused on continuously variable traits. He had proposed that evolution occurs as those heritable adaptations that confer a reproductive advantage are preferentially passed to successive generations, thus gradually becoming dominant within a species. But the consensus on natural selection, though relatively secure at the time of his death in 1882, broke down shortly thereafter. The reasons for this are complex. Certainly, there were widespread religious and moral objections to the materialism of Darwin's theory and haphazard nature of his evolutionary change. Yet humanist opponents of natural selection readily found scientific evidence in conflict with the theory. For example, the physicist William Thomson, from considerations of energy dissipation and heat conduction, calculated in 1868 an upper limit for the habitable period of the Earth that at around 100 million years, was far too short for natural selection to have taken effect.[3] And the form of inheritance proposed by Darwin would tend to blend adaptive modifications, thus countering the evolutionary effect of selection. Natural selection, it seemed, was insufficient to account for observed evolution; a more important mechanism must be at work.

Lamarck's thesis that acquired characters could be inherited by successive generations offered some prospect of progressive or directional evolution, and thus relief from the apparent purposelessness of Darwinism. Various forms of neo-Lamarckism were rampant by the late 1890s. Theodore Eimer popularized a related theory of 'orthogenesis,' in which straight-line evolution was driven from automatic forces inherent in the individual. Such theories

---

[3] Improved values for the conductivities of various types of rock reduced this value further. By the mid-1890s, Thomson was quoting a figure of 24 million years or less. Studies of rock strata and sediment thickness firmly indicated the Earth to be far older. Yet Thomson's prestige and the status of physical law as fundamental were widely regarded, at least at first, as more credible than the flimsier evidence of the amateurish geologists. More on Thomson and the age of the Earth in Chapter 4.

drew support from observations in the nascent field of embryology that seemed to show the genesis and development of the individual – ontogeny – as a recapitulation of that for a species – phylogeny.[4] The biometric school of Francis Galton, W.F.R. Weldon, and Karl Pearson had set out to demonstrate Darwin's theory with statistical evidence. But as Darwinism fell into disfavor, such methods came to be seen as mistaken and old-fashioned. In England, William Bateson, a former pupil of Weldon, was influenced in part by Galton's argument for constant racial types to propose that species did not evolve gradually, but form in discontinuous jumps. The position was strengthened in 1900 by the independent rediscovery by the European biologists Hugo de Vries, Carl Correns, and Erich von Tschermak of work by the Czech monk Gregor Mendel. Some thirty-five years earlier, Mendel had published a study of discontinuous traits found in hybrid crosses of peas. He explained the inheritance of these traits in terms of the random distribution of discrete, heritable factors that passed unaltered from one generation to the next.

Notwithstanding the conservative nature of blending inheritance, and Galton's own views on the fixity of the species, the biometricians had had some success in demonstrating that natural selection could give rise to permanent evolutionary shifts. Pearson had shown mathematically that offspring regressed to the parental, not the ancestral, mean, and that if this mean were altered by selection, the distribution of a trait in successive generations could gradually be shifted one way or the other. Weldon demonstrated this cumulative effect in 1895 with his crab measurements. The Mendelians, however, countered that such effects were small and limited in scope. The key to large-scale evolution was instead discontinuous variation, caused by the introduction of new heritable factors. Only then could evolution be rapid enough to fit with Thomson's calculations of the age of the Earth. Both Bateson and de Vries suggested that such new factors arose from mutation, although their mechanisms were slightly different. (In de Vries's neo-Lamarckian version the mutations were effected by changes in the environment.) A series of 'genetic' experiments – Bateson coined this sense of the word – following the rediscovery of Mendel's work seemed to demonstrate particulate and discontinuous inheritance. Sir Archibald Garrod showed in 1908 that Mendelian

----

[4] Lamarckism was not fully discredited until after World War I. In the early 1920s, the Austrian zoologist Paul Kammerer alleged that nuptial pads in midwife toads were direct evidence of the inheritance of acquired characters. Kammerer's specimens, however, were found to have been doctored with India ink. Though Kammerer shot himself in 1926, shortly after the discovery of the fraud, it is now thought that he probably had no personal involvement in the chicanery, but was instead the hapless victim of an anonymous ill-wisher. See Koestler 1971, Gould 1980.

theory could account precisely for the inheritance of phenylketonuria, a disease that leaves the body unable to break down the amino acid phenylalanine, resulting in skin disorders and mental problems.

The Mendelians regarded their emphasis on the direct physiological causes of inheritance in individuals as contrasting favorably with the restriction of the amateurish biometricians to the gross data of coefficients and correlations between groups. Biology was not merely curve-fitting. But this criticism carried little weight with the positivist Pearson. On the contrary, he argued that scientific inquiry could be nothing more than the collection of correlations. The Mendelians' search for ultimate causes was spurious, and their 'genes' metaphysical. The evidence for continuous variation had been accumulating for many years, and could not be ignored. The biometricians pointed to the hopelessly crude combinatorial analysis of the Mendelians, and charged as anti-Darwinian their focus on individuals rather than regularities in groups. The biometricians also argued that traits that supposedly Mendelized were not really discrete at all. It was better to measure traits and do the analysis properly.[5]

The debate between the Mendelians and biometricians was acrimonious, and personal attacks during their frequent and heated public spats were the order of the day. During the 1904 meeting in Cambridge of the British Association, Bateson flung down in front of the audience a stack of back issues of *Biometrika*, the biometricians' journal, declaring it to be worthless. Pearson claimed that the repeated assaults on Weldon's character and integrity drove him to an early grave at age forty-six. Certainly the breeding evidence alone was not unambiguous: the most persuasive examples in each camp were simply dismissed as special cases by the other. Yet the depth of the hostility suggests that wider issues were at stake. In part, this was a dispute over disciplinary skills and territory: should evolution be studied by mathematicians or biologists? Donald MacKenzie has argued that more private issues were also to the fore. To what extent should the theory of evolution be applied to human populations and used to guide social policy? Pearson believed that a vigorous eugenic program was necessary if the race were not to degenerate into gibbering senescence. Such eugenic ideas were not uncommon in liberal and intellectual circles of the time. They were often born of a philanthropic sense of the unfairness of the class system. Yet the association with idealism and social radicalism was a bar to general acceptance. Further, the eugenic movement

---

[5] The evolutionary debate has the largest bibliography in the history of science. See, e.g., Coleman 1971, Allen 1975, Bowler 1984, 1992, Provine 1971, Desmond and Moore 1991. For the statistical side see, e.g., MacKenzie 1981, Gigerenzer et al. 1989, Krüger et al. 1987.

was threatened by the backlash against natural selection. Lamarckian inheritance could result in social progress without any need for Pearson's breeding policy. Thus the broadly conservative Bateson, though no friend of the working classes, opposed eugenic reform. Other Mendelians shared Pearson's commitment to eugenics, but differed widely in their recommendations for government action. Hence the ferocity and bitterness of biometricians' dispute with Mendelians. Both sides regarded the other as a genuine menace to society because they were perverting the case for eugenic action. The question of inheritance had to be resolved if eugenics was to make any headway. The adoption of an erroneous model – or indeed, public bickering between the camps – could wreck the prospects of eugenic reform completely.[6]

Cambridge was at the center of the debate. Bateson had been appointed to a university chair of biology the year before R.A. Fisher's arrival in 1909, and R.C. Punnett, his co-worker, was based at Fisher's college, Caius.[7] Yet Fisher, unburdened with the ideological baggage of the previous generation, could see the merits of both sides. The new genetic experiments were certainly exciting, but so too was Pearson's application of rigorous mathematics to the question of evolution. Fisher addressed the newly-formed Cambridge University Eugenics Society in late 1911 on Mendelism and biometry. In his view there was no conflict. The theories were complementary. It was clear also that since both relied on statistical notions – biometrics on the normal distribution found to describe many traits, and genetics on the combinations and permutations of distributing genes – some improvement in mathematical statistics was the route to reconciliation.

3.3. FISHER'S EARLY WORK

Fisher set about the literature with enthusiasm. Like Pearson, he saw that the study of inheritance to be persuasive must be mathematically rigorous and impartial. Yet the usual standard deviations, correlation coefficients, and significance tests developed by the biometric school strictly applied only to large quantities of data, and Fisher found little on the analogous distributions for small samples. An exception was the work of W.S. Gosset, who published

---

[6] See MacKenzie 1981, Kevles 1985. Since increasing numbers of social as well as physical variables seemed biologically determined, some direct modification through selection or programs of differential breeding appeared possible. Yet there was no easy relationship between theories of inheritance and particular attitudes to eugenics. Many Mendelians, in America especially, were fervent eugenicists.

[7] Punnett published a popular book on Mendelism in 1911, and took the first chair of genetics at Cambridge the following year.

under the pseudonym 'Student.' A statistician at the Guinness Brewery, who generated only limited quantities of data in his studies of the growth of barley, Gosset had presented in 1908 the z-distribution, later renamed the t-distribution, the first exact solution of a significance test for a small sample.[8] Though it followed a year spent at University College studying with Pearson, this work met with little interest from the biometricians. They regarded small-sample statistics as not only mathematically complex, but irrelevant. Statistical methods were intended to demonstrate the patterns of heredity and reveal natural selection in action. Small or ambiguous samples were useless: torturous mathematics would hardly be persuasive to Darwin's opponents; it was better simply to gather a larger sample. But Fisher saw the work as a mathematical challenge. At Stratton's instigation, he started to write directly to Pearson and Gosset with comments and queries on their earlier papers.[9] From the start, Fisher's energy and mathematical precocity overwhelmed the older men. A bewildered Gosset referred to facing reams of "foolscap pages covered with mathematics of the deepest dye."[10]

Fisher's first scientific paper, on the fitting of frequency curves, was published while he was still an undergraduate. He pointed out arbitrariness in both the least squares and Pearson's minimum χ^2 methods of estimation. Least squares was dependent on the scale of the axes, χ^2 on the grouping of data. But Fisher did more than criticize earlier work. He suggested an 'absolute criterion' – a direct and invariant procedure of maximizing a function proportional to the chance of a given set of observations occurring. Subsequently, he returned to the question of small samples. He applied the ingenious idea of using points in n-dimensional space to represent samples of size n – borrowed from Gibbs's formulation of statistical physics – to prove Gosset's z distribution for small samples. He used the same idea in a publication of 1915 to derive the exact distribution of the correlation coefficient.[11]

Fisher was also working to reconcile genetics and biometrics. Introducing the method of componential analysis in order to partition biological variance into heritable and non-heritable causes, he was able to ascribe 95% of the

---

[8] For large samples, ±2 standard deviations from the mean capture 95% of the population, but with only two data points, a range of ±12 is required.

[9] Although he had written on statistical methods for agriculture – through which he had encountered Gosset – Stratton's statistical work was quite distinct from Fisher's, chiefly an extension of astronomical error theory. His encouragement of Fisher was due more to general enthusiasm for the work of his student rather than any specific statistical interest.

[10] Gosset to Karl Pearson, 12 Sept. 1912. Quoted in Box 1978, p. 72; also Pearson 1968, p. 55.

[11] Fisher 1912; Fisher 1915. See Porter 1986, pp. 315–6.

variation to heredity and only 5% to the environment. The new analysis was powerful. Fisher followed an earlier suggestion by the statistician George Udny Yule, another of Pearson's colleagues, that the continuous variation observed by the biometricians could be described by an aggregate of many discrete genetic variations. He showed not only that Mendel's genetic theory could then give rise to the observed correlations, but that this was the only way in which certain observed features – such as the high sibling correlation relative to that between parents and offspring – could be explained.[12]

This promising start was interrupted by a difficult and unhappy period. The brilliant young scholar was rejected for active service at the start of World War I because of his chronically bad eyesight, and as his friends joined up left and Cambridge, he found himself unappreciated and adrift. A brief period as statistician to a London investment company gave way to various short school teaching jobs for which, impatient and socially maladroit, he was temperamentally unfitted. The President of the Eugenic Society, Major Leonard Darwin, fourth surviving son of Charles, had been impressed by Fisher's enthusiasm and academic abilities when he had addressed the society in 1912. But even Darwin's provision in 1916 of a paid day each week at the Eugenic Society, mostly spent reviewing books, left Fisher's financial position precarious.

Matters started to improve at the end of 1916. Fisher had developed an attachment to farm work during a trip to Canada in 1913, and while still engaged in one of his stints as schoolteacher found support, chiefly from the intellectual and independent wife of a college friend, to start a farm of his own. Though hard and dirty, outdoor labor is not without a certain rude dignity. Farming was also, Fisher declared, a uniquely eugenic profession: the middle-class duty of a large family was a positive advantage in reducing the workload.[13] The farm gave Fisher the opportunity to combine his continuing theoretical work with direct experimentation. He carried out genetic crosses, and observed directly the eugenic effects of selective breeding in the varying weights of piglets and egg-laying yield of hens. It also led to an upturn in his personal fortunes. In 1917, Fisher married Ruth Guinness, the sister of his farm project partner, and daughter of Dr Henry Grattan Guinness, an

---

[12] The advantage of using the 'variance' – Fisher coined this as a specialist term – is that, as the square of the standard deviation, it can be additively partitioned. Fisher was probably unaware of Edgeworth's early work on the analysis of variance. See Stigler 1986, pp. 312–14; Porter 1986, p. 268.

[13] Fisher had regarded the war was a eugenic disaster: the elderly and infirm remained safe while a good fraction of the able youth was exterminated. He also shared the popular middle-class perception that the rich and poor had shirked their responsibilities.

evangelical preacher. They set about the farm work and personal eugenic duties with fervor. Though impoverished, Fisher was happy.

3.4. THE CLASH WITH PEARSON

Fisher's paper of 1915 had not gone unnoticed. The correlation coefficient was dear to Pearson, and an exact distribution had eluded the biometricians for years. His response appeared two years later in a 'cooperative study,' which tested the accuracy of Fisher's supposedly exact distributions against previously-published approximations.[14] Pearson chided the young man. His derivation, like his 'absolute criterion' of 1912, was a misapplication of inverse probability. Pearson was of course committed to inverse probability. But his objection to Fisher's analysis was not that of the Bayesian offended by the novelty of frequentist methods. On the contrary, Pearson took the 1915 paper as a straight application of inverse probability. Instead, his target was Fisher's supposed use of the Principle of Insufficient Reason. The "equal distribution of ignorance" had "academic rather than practical value"; to be applicable, equal prior probabilities for the values of certain parameters could only be justified if these parameters occurred with roughly equal frequencies in large-scale statistical studies. To demonstrate, the cooperators proceeded to use non-uniform priors based on surveys of the parent-child correlation. They ended their paper with a mild rebuke: the Principle of Insufficient Reason "has unfortunately been made into a fetish by certain purely mathematical writers on the theory of probability, who have not adequately appreciated the limits of Edgeworth's justification of the theorem by appealing to general experience."[15]

Perhaps the cooperative study was to be a gentle reminder that the eager young man – and potential rival – should be a little more respectful to his elders. Or, with the cooperative study billed as an appendix to papers from both Fisher and Student, perhaps Pearson intended his words for Gosset, as a continuation of a recent exchange over the choice of uniform priors for

---

[14] Soper et al. 1917.

[15] "Bayes' theorem ought only to be used where we have in past experience, as for example in the case of probabilities and other statistical ratios, met with every admissible value with roughly equal frequency. There is no such experience in [Fisher's] case." Soper et al. 1917, p. 358. Pearson usually appealed in his work, however, to a very general sort of experience to license the Principle of Insufficient Reason: since "in cases where we are ignorant, there in the long run all constitutions will be found to be equally probable," so could the principle be tentatively applied even in situations where no directly relevant knowledge was available. This was justified with an appeal to Edgeworth's empirical studies. (Pearson [1911], p. 146.)

the standard deviation of a normal population.[16] In any case, Pearson reckoned without Fisher's sensitivity. Although starting to make a name in the field, Fisher was still professionally vulnerable, and his personal circumstances were hardly more secure. On reading the paper, Fisher was indignant. Pearson's criticisms were unjust and ill-directed; he had not even had the courtesy to warn Fisher that such a reply was in preparation.[17] Reviewing his correspondence with Pearson, which had run cordially since 1912, Fisher discerned snubs. A note he had sent in 1916 pointing out the arbitrariness of the minimum χ^2 method to grouping of data had gone unpublished. And the paper on Mendelism, completed in the summer of 1916, had also been rejected. Fisher detected Pearson's hand at work.[18]

Though constitutionally volatile and over-sensitive to criticism, Fisher held himself in check. But he was resentful on behalf of Leonard Darwin, by now something of a mentor, whose work on the influence of the environment Pearson had criticized, and made a vain attempt to persuade the pacific and genial Darwin to respond. The clearest evidence of Fisher's rancor followed Pearson's offer in 1919 of a position as chief statistician at the Galton Laboratory. This was a unique opportunity for professional recognition. The laboratory had been founded at University College London in 1904 with the sum of £1,500, plus an annual endowment of £500, donated by the childless Galton; on his death in 1911, his will provided for Pearson's chair in eugenics. It was the only place where statistics was taught or studied to any depth, and was already attracting students from around the world. A post at the Galton Laboratory was both prestigious and a step to academic security. Yet though Pearson's offer followed a number of academic rejections, even from posts in New Zealand and Cairo, Fisher regarded it with suspicion. At around the same time, Fisher received a second offer, to fill the newly-created post of statistician at the Rothamsted agricultural experimental station. Given his own agricultural work, this was not an entirely unattractive alternative. Deciding the duties at the Galton would be too restrictive, and doubting his

[16] See Aldrich 1995.

[17] See Pearson 1968, Edwards 1974, Box 1978, p. 79. Fisher apparently heard of the cooperators' paper only when it was in proof.

[18] In the case of the paper on Mendelism, Fisher was not mistaken. With disciplinary authority and eugenic issues at stake, neither the biometricians nor the Mendelians were willing have their dispute settled. Pearson had objected to Yule's original argument that continuous variation was the limiting case when many discrete factors were acting, and was no more favorably inclined toward Fisher's more mathematical version. As referees, he and Punnett rejected the paper for the Royal Society. Darwin's influence and financial support were needed to get the work published. It finally appeared in late 1918 in the *Transactions of the Royal Society of Edinburgh*.

freedom to publish independently from Pearson, he turned his offer down, and instead moved to Rothamsted.

3.5.1. Fisher's new version of probability

The Experimental Station at Rothamsted had been established in 1837 to study the effects of nutrition and soil types on plant fertility. Although the research had lost direction toward the end of the nineteenth century, parliamentary reforms from 1909 promised a boost for scientific agriculture. A new director, John Russell, was anxious that several decades' worth of accumulated data be not wasted. Analyzing this was to be Fisher's task.[19] Although initially hired for a six-month period, Fisher's expertise was soon recognized. Russell decided that he was a genius who must be retained.

Installed at Rothamsted, Fisher had the academic freedom to reply to the cooperators' study. First, he needed to reexamine his concept of probability. Fisher had learned Inverse Probability as standard at school and "for some years found no reason to question its validity."[20] The papers of 1912 and 1915, though technically innovative, had traded on the old epistemic interpretation. Yet Fisher was aware of the limitations of the theory and the existence of the rival version. Dr John Venn, the popularizer if not originator of the frequency definition was, after all, the President of Caius when Fisher was an undergraduate. The cooperators' attack was the occasion for careful consideration of the two probability models.

Fisher's work on genetics seemed relevant. Mendel's was one of the first wholly probabilistic theories in natural science. As in the case of Brownian motion, or the velocity distributions of statistical physics, Mendelian probabilities were perhaps merely descriptive conveniences, a summary of the long-term behavior of an underlying mechanism, rather than the consequence of stochastic indeterminacy at the atomic level. Yet Mendelism was unique in involving a chance mechanism that generated with exact and fixed probability one of a set of clearly-defined outcomes. Genetic probabilities could thus be *treated* as inherent to the world rather than reflecting incomplete knowledge. As George Barnard has written, Mendel's theory was the first indication that "nature itself could behave like a perfect gambling machine."[21] Randomness

[19] Box 1978, p. 95.
[20] Fisher 1936, p. 248.
[21] Barnard 1989, pp. 259–60; Barnard 1990.

explained the observed outcome, rather than requiring itself explanation in terms of underlying causes.[22] Mendelism, like throws of a die or tosses of a coin, calls for a frequency definition of probability. By definition, gametes distribute by chance, and the long-run frequency of a given genotype in a large offspring generation can be predicted exactly from the genetic make-up of the parents and the rules of combinatorial analysis. But if probabilities referred only to events, how could the credibility of a hypothesis be assessed? Again, Mendelian genetics provided a suggestion. The routine form of pedigree analysis for Mendelian crosses produced a ratio that provided a measure of a hypothesis's credibility, while being non-additive and thus not a probability. Fisher began to see that there was more than one measure of uncertainty.[23]

3.5.2. The papers of 1921 and 1922

Fisher developed his new version of probability in his response to the cooperators. Though rejected in 1920 by Pearson's *Biometrika*,[24] the paper finally appeared in the journal *Metron* in 1921. Leonard Darwin had advised him to tone down an earlier draft, but the published paper was still strong stuff. Fisher rejected totally not only 'Bayes's postulate' – the Principle of Insufficient Reason – but Pearson's whole epistemic interpretation of inverse probability. Both were "widely discredited"; Fisher had used neither in his own work. Further, the cooperators had criticized a non-existent use of the Principle of Insufficient Reason, only to advocate its replacement by an equally arbitrary *a priori* assumption. This blunder they compounded with elementary errors in the mathematics.

In fact, Fisher countered, it was the cooperators who did not understand probability. They applied the concept to both events and hypotheses. Yet these are distinct things, each with a separate measure of uncertainty. To insist that a measure of rational belief should obey the addition rules of probability, as did the inverse probabilists, was an unnecessary restriction. We use probability just for observations or events, and something different, which Fisher dubbed

---

[22] Chance was 'causal' in Boltzmann's statistical physics, in that atomic-level randomness gave rise to large-scale regularity. And chance appeared in Darwin's theory as the origin of inherited variation (although the successful propagation of such variation depended on adaptive advantages bestowed on the host organism). For both, though, these 'chances' were rather epistemic probabilities, a consequence of imperfect knowledge of events governed in nature by deterministic physical laws. (The probabilistic status of radioactive decay was still to be settled. See von Plato 1994, pp. 137–41.)

[23] See Fisher 1918.

[24] Fisher read the note accompanying the rejected paper as uncivil, and vowed never again to submit to Pearson's journal. He didn't. Box 1978, Pearson 1968, Bennett 1983, p. 73.

'likelihood,' and which was like the ratio in Mendelian pedigree analysis, to express our confidence in theories or hypotheses.[25]

This, declared Fisher, had been the point of his 'absolute criteria' and the 1915 paper. He had derived his expression for the exact distribution of the correlation coefficient with what he now proposed to call the Method of Maximum Likelihood. Although superficially and numerically similar to an application of inverse probability with the assumption of uniform priors – "equal distribution of ignorance" – the method was conceptually distinct. It finds the value of the parameter, called by Fisher the 'optimum,' that gives the highest chance to the observations actually obtained. Unlike the inverse probability estimate, this optimum value is absolute, and would follow from a different transformation of the correlation.[26]

What precisely, then, are probabilities? Fisher sharpened his position in a major paper published the following year in *Philosophical Transactions of the Royal Society*. Using the abstract terms of a new statistical model, he defined a probability explicitly as a frequency ratio in a 'hypothetical infinite population,' from which the actual data are a random sample.[27] Fisher argued that the practice of biometricians from Galton onwards of extrapolating probabilities from huge quantities of data already enshrined the frequency notion. The infinite hypothetical population definition simply made this practice explicit, and enabled probabilities to be calculated for smaller samples. The inverse probabilists confuse probabilities of the sample with those of the parent population.

[25] "We can discuss the probabilities of occurrences of quantities which can be observed or deduced from observation, in relation to any hypotheses which may be suggested to explain these observations. We can know nothing of the probability of hypotheses or hypothetical quantities. On the other hand we may ascertain the likelihood of hypotheses and hypothetical quantities by calculations of observations; while to speak of the likelihood (as here defined) of an observable quantity has no meaning." Fisher 1921, p. 25.

[26] Measures similar to Fisher's likelihood had been used both for point estimation, usually in error analysis, and as numerical assessments of the truth of a hypothesis long before 1922, and by authors as diverse as Bernoulli, De Moivre, Laplace, Lagrange, Gauss, and even Keynes. In his discussion of the history of likelihood, Edwards traces the first likelihood argument to Daniel Bernoulli in 1777. The astronomer Lambert argued in a similar direction around the same time. Often such applications depended directly on an assumption of Bayesian priors, and were not intended as 'best' methods in the sense of Fisher's later specialized terms. Fisher's paper of 1922 was the first to clearly specify the non-Bayesian approach and to discuss properties of the estimators. It presented the method of maximum likelihood as a wholly independent method of support for hypotheses. See Edwards 1972, 1974.

[27] The population is 'hypothetical' in the sense that it is infinite and hence unrealizable, and that repeated sampling from it might prove practically impossible. The variation in natural conditions introduces a different set of causes to a repeated agricultural experiment, for example, so changing the relevant super-population.

This was "strikingly exemplified" by a recent paper of Pearson. But since we can generally say nothing about the frequencies of parameters of a population from a sample, the frequency concept is mathematically inadequate to measure the credibility of a hypothesis respecting these parameters. Probabilities are used solely to calculate sample distributions from populations. It is likelihood, not probability, that properly expresses confidence in the population given the sample.[28]

Fisher illustrated the point with a Mendelian pedigree test. Coat color in mice is controlled by a single gene, which comes in two forms – called alleles – labeled as black and brown. The black allele is dominant to the brown. That is, if either or both of the alleles inherited from the parents are black, the mouse will have black fur; only if both alleles are brown will it have brown fur. We wish to test whether a particular black mouse has different or the same alleles; that is, whether the mouse is heterozygous or homozygous. We mate the mouse with a brown mouse and observe the colors of the offspring. The probabilities of our particular result – seven back offspring – given the proposition that the original mouse was homozygous and heterozygous are 1 and $\frac{1}{128}$, respectively. Bayes's theorem would lead to a posterior probability of heterozygosity on the evidence, but would depend on whatever prior probability we gave to the original mouse being heterozygous. Knowing something about the lineage of this mouse, we could assess this prior probability, and hence the posterior probability, in frequency terms. Otherwise, "no experimenter would feel he had warrant for arguing as if he knew that of which in fact he was ignorant." The likelihood ratio, however, is a perfectly objective $\frac{128}{1}$.[29]

To promote his new interpretation of statistics, it was necessary for Fisher to refute the rival of inverse probability. He took to the task with relish. His chief objection was to the arbitrariness and inconsistency of the Principle of Insufficient Reason. "Apart from evolving a vitally important piece of knowledge out of ignorance, it is not even a unique solution." He did not trouble to examine the relative merits of epistemic and frequency probabilities, but

---

[28] Fisher wrote: "in an important class of cases the likelihood may be held to measure the degree of our rational belief in a conclusion." Fisher 1922a, footnote to p. 327. See also Fisher 1930a, p. 532: it provides "a numerical measure of rational belief, and for that reason is called the *likelihood*." Likelihood is not probability: "If A and B are mutually exclusive possibilities the probability of 'A or B' is the sum of the probabilities of A and B, but the likelihood of A or B means no more than 'the stature of Jackson or Johnson'; you do not know what it is until you know which is meant. I stress this because in spite of the emphasis that I have always laid upon the difference between probability and likelihood there is still a tendency to treat likelihood as though it were a sort of probability." Fisher 1930a, p. 532.

[29] See Barnard 1989, p. 259.

instead recast the general experimental situation in explicitly frequentist terms. If the population of interest is itself drawn from a known super-population, we can deduce using perfectly direct methods the probability of a given population and hence of the sample.[30] But if we do not know the function specifying the super-population, we are hardly justified in simply taking it to be constant. Not only is this choice of *a priori* distribution completely baseless, but the restatement of our population using different parameters would in general lead to a different function. A prior probability distribution can only be verified by sampling from reference set. Yet the Principle of Insufficient Reason had been habitually misapplied to general cases and hypotheses. Indeed, announced Fisher, its indiscriminate use by earlier writers, such as Gauss on the method of least squares, contaminated the language of statistical inference and vitiated most of their work.[31] More recently, it had been the biometricians' confusion between a sample and the population from which it was drawn that had allowed "the survival to the present day of the fundamental paradox of inverse probability which like an impenetrable jungle arrests progress towards precision of scientific concepts."[32]

3.5.3. The Pearson–Fisher feud

Joan Box, Fisher's daughter and biographer, describes a "fluttering in statistical dovecotes" as the community awaited Pearson's response to this dense and novel work. It came obliquely in *Biometrika,* as a 'further note' to a short paper Fisher published also in 1922 concerning Pearson's use of the χ^2 test.[33] Pearson was in combative mood: Fisher was wrong and obviously wrong. Moreover, he "had done no service to the science of statistics by giving [his mistakes] broadest publication" in the Journal of the Royal Statistical Society. He "must either destroy himself or the whole theory of probable errors..."

---

[30] Fisher 1922a, p. 325. This joint probability could then be integrated over all values specifying the population to give the probability that these specifying parameters lie within any assigned limits.

[31] Bayes himself was not one of the guilty men. Fisher presented Bayes's thinking in frequency terms ("He imagines, in effect, that the possible types of population have themselves been drawn, as samples, from a super-population..."), and regarded Bayes's tentative choice of the billiard table example – for which Bayes's Postulate is justified – and his reluctance to publish as evidence of doubt over the more general application of the theory. It was Laplace who had caused all the problems, by "cruelly twisting the definition of probability itself in order to accommodate the doubtful axiom" (Fisher 1936, p. 247; see also footnote 33, chapter 2). Bayes is the second most cited author in Fisher's collected papers.

[32] Fisher 1922a, p. 311.

[33] Pearson 1922.

Fisher rejoined immediately, but the publishing committee of the Royal Statistical Society, nervous after Pearson's comments, turned his paper down. Darwin's soothing words, and advice that any action would win him a reputation of awkwardness, were ignored, and Fisher resigned from the society. In the next few years, and under the guise of replies to specific queries from other statisticians, he published a series of papers that extended and clarified his position.

Henceforward, Fisher and Pearson were implacable opponents. Exacerbated by personal animosity – Pearson's freethinking rankled with the Anglican and innately conservative Fisher – and fueled by blunt and bruising polemic, the feud grew in intensity as the years wore on.[34] Fisher kept up a steady barrage and rarely missed an chance to either attack Pearson directly or snipe at his advocacy of Inverse Probability. Fisher had created many such opportunities in the 1922 paper. His failure to distinguish likelihood from probability in 1915 had precipitated the dispute with Pearson. Fisher would not make this mistake again. With a genius for loaded terms, he used his infinite population model to coin, in addition to 'optimum' and 'likelihood' from the 1921 paper, 'efficiency,' 'sufficiency,' and 'consistency' as specialist terms.[35] His method of maximum likelihood was the best route to the estimation of population parameters, since it would always give rise to a statistic that was efficient (and sufficient, too, if one existed).[36] The mean square error was sufficient, and thus "an ideal measure of the parameter." Pearson's methods, on the other hand, were inefficient, insufficient, or inconsistent. The method of moments, for example, though at the very least non-arbitrary was inefficient unless the frequency curve is nearly normal. And minimum χ^2 was valuable only because it was a large sample approximation to maximum likelihood. Fisher hammered home the attack in a series of papers on the

[34] Nor was a truce called when Pearson died. Twenty years later, Fisher spoke of Pearson running his lab like a sweat shop and ill-treating his assistants, and castigated his lumpen mathematics and 'pretentious and erratic' biological ideas. "The terrible weakness of his mathematical and scientific work flowed from his incapacity in self-criticism, and his unwillingness to admit the possibility that he had anything to learn from others, even in biology, of which he knew very little. His mathematics, consequently, though always vigorous, were usually clumsy, and often misleading. In controversy, to which he was much addicted, he constantly showed himself to be without a sense of justice" (Fisher 1956, p. 3).

[35] An efficient statistic has the least probable error in those situations where it tends to normality; a sufficient statistic summarizes all relevant information contained in the sample; a consistent statistic, when calculated from the whole population, is equal to the required parameter. These terms and concepts had been applied in the field before, but Fisher was the first to define them precisely, and to tie them into a coherent statistical model.

[36] Fisher did not prove this rigorously in 1922.

χ^2 test published during the twenties. He cleared up long-standing problems with the test by proving that the number of 'degrees of freedom' – another coinage – needed to be reduced by the number of linear constraints on the data. Not only did Pearson not understand his own creation, he suggested, a large part of his statistical work was nugatory.[37]

Despite their differences over Mendelism, Fisher and Pearson shared eugenic goals and a commitment to transform statistics into a standard tool for scientific practice.[38] Yet even in the early papers, Fisher had signaled that the biometricians' good intentions did not excuse their disorganized and inconsistent methods. (He had lamented the simultaneous promotion of separate estimation procedures, including the method of moments, minimum χ^2, and the maximization of the posterior probability: the numerical results might be similar, but in obscuring the theoretical distinctions the biometricians were doing no service to the discipline.[39]) As Fisher presented it, the debate was not over the numbers – although Pearson's misapplications of the χ^2 test could be consequential – but his attempt to reform the mathematical and logical foundations of the subject. For his part, though perhaps unprepared for the ferocity and relentlessness of Fisher's opposition, Pearson remained wedded to his positivist philosophy and his empirical version of Bayesianism.[40] (He later wrote: "I know I have been preaching this doctrine vainly in the wilderness for many years, and made a distinguished statistician a permanent enemy by suggesting it, but I believe it to be correct.") Many of the older generation, however, reluctantly began to concede that Fisher was right.

Do Fisher's papers of 1912 and 1915 share an essential basis with the sanitized and explicitly frequentist method of maximum likelihood he unveiled in 1922, or are these earlier works reliant on hidden Bayesian assumptions? The evolution of Fisher's concept of likelihood and his exchange with Pearson have been extensively analyzed. Some statisticians have attempted to apportion blame, others to reconstruct the debate along contemporary lines

[37] In 1925, Fisher wrote that Pearson's work on the χ^2 test, although marred by a "serious error, which vitiated most of the tests of goodness of fit made by this method until 1921," can nevertheless be salvaged as the distribution of the variance of normal samples (Fisher 1925b, p. 17). Pearson never accepted Fisher's criticisms of his test.

[38] Pearson had a number of inspiring texts painted up onto the walls of the Galton Laboratory, including Lord Goschen's "Go with me into the Study of Statistics and I will make you Enthusiasts in Statistics."

[39] Aldrich 1995 notes that textbooks of the time would give separate derivations for the standard error of the sample mean without either distinguishing them or indicating their conceptual differences.

[40] Pearson refused permission, possibly for financial reasons, for his χ^2 tables to be reproduced in Fisher's *Statistical Methods for Research Workers*.

in order to establish a respectable lineage for newer concerns.[41] The issue is confused by the imprecise and inconsistent use of terminology that prevailed at the time. Thus 'Inverse Probability,' 'Bayes's theorem,' 'Bayes's postulate,' and so on refer sometimes to the same, sometimes to different things. It is further complicated by the rather favorable gloss Fisher subsequently gave his earlier papers. (Despite later implications, the likelihood was not presented as the sole route to inference in the papers of 1912 and 1915, for example.) And it is perhaps irredeemably muddied by the entanglement of concepts from direct and inverse methods, a legacy of the distinct strands of probability theory in biometrics and error theory. Scholars are divided. In her biography, Joan Fisher Box states that Fisher had been perfectly clear and Pearson in error. Pearson's son Egon, in contrast, has written that Fisher's 1915 paper was muddled, and his father justified in assuming that the Principle of Insufficient Reason had been used: with space restricted in *Biometrika* due to war shortages and financial strain, Pearson had simply wished to avoid unnecessary controversy. More disinterested writers also differ. Zabell thinks that Fisher clearly avoided any Bayesian taints, while Aldrich declares that the 1912 paper was based on inverse probability, and that it was later, in 1921, that Fisher "quietly uncoupled" maximum likelihood from Bayesianism. Conniffe suggests Fisher was following Keynes; Yates and Mather that Pearson was simply closed-minded. Edwards and Barnard make the strong case that the framework for likelihood, if not the term itself, was present in 1912, and thus that the break with inverse probability, if not clean, was at least clear.[42]

In the early papers, Fisher certainly made plain that what he was later to call the likelihood was not a conventional probability. It was a relative rather than an absolute measure, and could be used "to compare point with point" but not added or integrated like a continuous probability. A likelihood measures whether parameters have specified values; a probability that these values lie between certain limits (i.e., fall within some differential element). Unlike a probability, the relative values of the likelihood are invariant under a

---

[41] Student and Laplace are often outed by neo-Bayesians, but was even Fisher a Bayesian? Certainly not in the usual sense. Yet Fisher believed that Bayes's interpretation of probability was based on expectations, and thus frequencies in a super-population, and regarded it as identical with the concept of fiducial probability, which he unveiled in 1930. (Fisher did not use word 'fiducial' as a modifier, but simply to make clear that it referred to events that have already occurred.) See Lindley 1990, Barnard 1987, Jeffreys 1974.

[42] Box 1978; Pearson 1968, 1974; Zabell 1989b, p. 261; Aldrich 1995; Conniffe 1992; Yates and Mather 1963; Edwards 1972; Edwards 1992, p. 98; Barnard 1963, 1990. See also Savage 1976.

transformation of scale.[43] Fisher did not make use of a Bayesian, or uniform prior. Yet the full population interpretation is absent in 1912. Despite his admiration for Student's small sample work, Fisher followed the biometric tradition in not fully distinguishing between parameters of the sample and statistics of the population.[44] And there is little evidence that he had renounced inverse probability before 1917. Indeed, though not using the word 'prior' in either the 1912 or 1915 papers, he does refer to "the inverse probability system" when discussing his method, and criticizes an alternative technique as having "no definite meaning with respect to inverse probability."[45] As late as 1916, in a note to Pearson, Fisher defended the method of moments for the normal curve against the charge of arbitrariness on the grounds that it flowed from the "Principle of Inverse Probability."[46] And Fisher confessed in his own 1922 paper that he "must indeed plead guilty in my original statement of the Method of Maximum Likelihood [of 1912] to having based my argument on the principle of inverse probability."[47] Thus though Pearson's criticisms were undoubtedly misplaced, his slip is understandable given Fisher's repeated talk of inverse probability, and can probably be chalked up to the time pressure on a busy man, or the unresponsiveness to conceptual innovation natural given the length of his involvement with the material, rather than deliberate malice.

Fisher did, however, take the cooperators' attack as an opportunity to revise his concept of probability. But although the exchange gave Fisher the impetus to rethink his interpretation, it was Mendel's theory that gave him the means clearly to distinguish epistemic and frequency probabilities, and reasons for preferring the latter. A probability, like the frequency of a gene, was naturally definable in terms of a ratio in a population. Prior probabilities were assessments of the distribution of these populations; in genetics, they could be generated objectively by combinatorial analysis of an individual's lineage. Bayes's theorem was not a mysterious algorithm for the scientific

---

[43] Fisher 1912, p. 160: The likelihood "is a relative probability only, suitable to compare point with point, but incapable of being interpreted as a probability distribution over a region, or of giving any estimate of absolute probability."

[44] In 1922, Fisher recommended that statistics of the population and parameters of the sample be denoted by lower-case Greek and Roman letters respectively.

[45] Fisher 1912, pp. 158, 159.

[46] See Pearson 1968 [1970], pp. 414–5.

[47] Fisher 1922a, p. 326. Fisher followed this quote by pointing out in mitigation that this did not mean that he had used the Principle of Insufficient Reason, and that he had at least made clear in 1912 that what he called 'probabilities' – now likelihoods – were quite distinct from the usual quantities referred to by that term, because they were relative rather than absolute.

method or the revision of beliefs in the light of new knowledge, but simply a logical relationship between direct frequencies of populations. Mendelian probabilities concerned the distribution of gametes; they did not change with new information. Mendelian practice also indicated that there were at least two ways to measure uncertainty. For Fisher, Mendelism provided not merely a model of probability. It also served as a model of standard science, in which hypotheses refer to repeatable experiments that concern distinct and independent outcomes, each clearly defined and of precise probability.[48] Such a model found ready interpretation at Rothamsted.

3.6. THE MOVE TO ROTHAMSTED: EXPERIMENTAL DESIGN

Fisher's perceptive analysis of crop data soon made an impact at Rothamsted, and experimentalists at the station began to seek him out during tea breaks for advice on practical problems. These agronomists were not content to summarize their data. New agricultural techniques were to be tested and assessed: a sustained increase in the yield and quality of the crops meant increased profits for farmers. Yet this was not a straightforward matter. Unlike a rigorously controlled experiment in a physics laboratory, the plurality of possible influences – ranging from the weather and the effects of birds, to differences in soil type and irrigation – and the wide variation inherent in natural populations made it generally impossible for an investigation of the effects of different fertilizers or plant strains to produce an unambiguous result.

The subject was made for statistical inference. The growth of plants and vegetables generates readily-quantifiable data, and the gestation of each generation is short enough to detect beneficial effects and thus improve farming practices. Yet although agricultural societies had been conducting experiments since the eighteenth century, they had lacked the mathematical tools to draw any but the most rudimentary conclusions. The mean yield of plots were often compared during the nineteenth century, but the importance of the sample size on the reliability of the conclusion was not generally appreciated; replication of a field experiment was valued only inasmuch as the difference between trials could be used as a rough estimate of the error of the results.[49]

The statistical techniques of the biometricians were more sophisticated, but were ill-suited to inference. Instead, the biometricians' emphasis on

[48] Barnard 1990, pp. 24, 27.
[49] See Swijtink in Grattan-Guinness 1994, pp. 1363–70; Gigerenzer et al. 1989, pp. 70–90.

description was reflected in their habitual conflation of the statistics of their empirical surveys with the parameters of the underlying population.[50] But as Fisher had pointed out in 1922, these quantities were distinct, and the problem of estimating one from the other a matter of inference. Fisher had no truck with Pearson's positivist belief that speculation about causes was worthless. Inference, as much as description, was a job for the statistician. The problem was the sample size. The larger the sample, the better the approximation to the complete population. But agricultural data were less tractable than the social measures of the biometricians. The number of complicating factors in a field experiment make even a large crop relatively small for inferential purposes. Thus Fisher regarded small sample work not merely as a chance to flex his analytical muscles – it had prompted the mathematically-involved proof he had sent to Student in 1912 – but as a route to statistical inference. He had continued working on exact sampling distributions for small sample statistics through 1915, and extended these methods at Rothamsted during the twenties.

There was another side to the problem of inference. As the quantity of information decreases, so its quality becomes more important. Some sorts of data are more revealing than others. Fisher began to see that squeezing information from small samples would be easier if those samples were somehow well behaved, or suited to particular analysis. Rothamsted was a good base to consider these questions, and Fisher started to think about how best to design experiments to yield good-quality data. There was much folk wisdom in the standard practices of agriculturists, but it lacked a formal mathematical development. Fisher borrowed the analysis of variance from his earlier genetics paper. Several potentially causal factors could be investigated simultaneously by dividing the experimental plot into units reflecting all combinations of factors; unanticipated effects such as fertility gradients could be balanced by comparing only pairs treated identically save for one factor.[51] Then the variance could be partitioned into different causal and error components and compared with control groups to test for efficacy of treatment. Fisher also introduced the requirement of randomization. Subconscious selection or questionable assumptions of independence could be eliminated if the treatment received by each unit were chosen at random. The idea is counter-intuitive.

---

[50] The conflation of values of specific parameters of a model with a measure of the goodness of fit of that model can also be glossed as a confusion between tests of estimation and significance. As mentioned, Fisher blamed this for the continuing survival of inverse probability.

[51] A plot divided as a 'Latin Square' – an $n \times n$ array of n different symbols in which each appears exactly once in every row and column – has the desired properties.

To draw experimental conclusions, surely we should remove random scatter, not introduce more? Shouldn't a set of units carefully selected from widely dispersed parts of the plot lead to more representative results than a random selection that might by chance turn out to be from the same area and thus skewed by very local effects? Yet Fisher showed such arguments to be misguided. Randomization was the only route to valid significance tests, since it was only when the treatment assigned to each unit was chosen at random that additional biases due to causal influences either unsuspected or uncontrolled could be eliminated.[52] Fisher intended his new methods to standardize the multifarious variation of nature. The agricultural plot was to become the controlled environment of a physics laboratory. In this way, experiments were defined to yield clear and unambiguous answers. Though loath to concede wholly the skill of the experimenter to statistical design, the agronomists at Rothamsted were impressed.

3.7. THE POSITION IN 1925 — *STATISTICAL METHODS FOR RESEARCH WORKERS*

Fisher's book, *Statistical Methods for Research Workers*, is a snapshot of his position in 1925. It combines the insights of experimental design developed at Rothamsted with the conceptual statistical framework of the 1922 paper. How is an experiment to be analyzed? First, the experimenter approaches the situation with particular questions to be answered. Based on these questions, and pragmatic considerations, he constructs an experimental model and outlines a hypothesis to be tested. The assumptions of the model specify a unique hypothetical infinite population. Second, he determines suitable test statistics, and uses their values as calculated from the data as estimates of the parameters of the population. Third, he derives distributions across random samples from the population of these estimates of the parameters, and of other statistics designed to test the specification.[53]

The analysis is a mix of inductive and deductive methods, and addresses distinct questions. Does this treatment work? If so, by how much? When

---

[52] Random here does not mean arbitrary or haphazard, but according to some chance-like process such as the toss of a coin. The estimate of errors, and hence test of significance, made for randomized plots is valid because "if we imagine a large number of different results obtained by different random arrangements, the ratio of the real to the estimated error, calculated afresh for each of these arrangements, will be actually distributed in the theoretical distribution by which the significance of the result is tested." Fisher 1926, p. 507. Though still contentious, randomization is often held to be the greatest of Fisher's many innovations. For his theory of experimental design, see Fisher 1926, 1935a.

[53] Fisher summarized these steps as specification, estimation, and distribution.

the form of the population is known, its parameters can be estimated from the data directly, and the uncertainty in these estimates expressed in terms of a likelihood function. In more doubtful situations, a weaker form of inference, the significance test, can be used to assess the form of the population. In these tests the population is usually specified by a 'null hypothesis,' defining the lack of a measurable effect. Then the 'significance level,' P, is calculated – the probability that the test statistic equals or exceeds the value actually obtained in the experiment, assuming the null hypothesis is correct. Now P cannot of course be interpreted as the probability that the null hypothesis is true. But if P *is* small, we can say that either we have obtained an unlikely set of data, or the null hypothesis is false. "If P is between .1 and .9 there is certainly no reason to suspect the hypothesis tested. If it is below .02 it is strongly indicated that the hypothesis fails to account for the whole of the facts. We shall not often be astray if we draw a conventional line at .05, and consider that [lower values] indicate a real discrepancy."[54] P can also be interpreted as a kind of error probability, telling us how often a result as or more extreme than the one observed would occur if the null hypothesis is true. "Either there is something in the treatment, or a coincidence has occurred such that does not occur more than once every twenty trials."[55]

As a statistical hypothesis, the null hypothesis can never be definitively either proved or disproved. Fisher believed, however, that the significance test could assess its 'adequacy' as a specification of the population, and could in certain circumstances justify its rejection as 'contradicted' by the data. Since the distribution of sample statistics can be derived before any data is recorded, Fisher's recipe could be followed to plan the experiment so that the data obtained would be sufficient to test the hypothesis to the desired level of reliability and significance. Thus are design and analysis linked.

Much of the new material in *Statistical Methods for Research Workers* was developed specifically in response to practical problems of research, and most of Fisher's illustrations came from agriculture and genetics – for example, how to test the exact ratio predictions of a Mendelian hypothesis with the real data, subject to fluctuations, from a breeding experiment. Yet as Fisher made clear, statistics is not merely a branch of these or other studies. The same few categories of distributions are applicable in many, seemingly distinct areas. Fisher presented statistics as a general technique for the mathematical treatment of data, a sort of mathematical expression

[54] Fisher 1925b, p. 79.
[55] Fisher 1926, p. 504.

of proper and objective scientific practice. The technique could be generally applied: "Statistical methods are essential to social studies, and it is principally by aid of such methods that these studies may be raised to the rank of sciences."[56]

Fisher's outline, however, is of more than an underlying method. First, in combining design with analysis, he prescribes not merely analytical method, but proper experimental conduct. His outline thus carries a moral imperative, and is part of a long tradition – dating to the early applications of error theory in astronomy – of tying objectivity of results to standardization of experimental practice, and thus socialization of the individual.[57] Second, Fisher's recipe for good scientific practice, free from the arbitrariness of inverse probability, is for a specific sort of science. This has as its standard the Mendelian model: nature as gambling machine, generating data of repeatable and independent trials. Third, statistics concerns populations rather than individuals: as Fisher made clear, the practice of repeated measurement "shows a tacit appreciation of the fact that the object of our study is not the individual result, but the population of possibilities of which we do our best to make our experiments representative."[58] Good sciences, such as the kinetic theory, involve statistical aggregates and representative experiments.

The emphasis on design and standardization was new, and apart from that of Student – by this stage a friend of Fisher – the initial reviews, from the biometricians who still dominated the statistical community, were uniformly negative. Small samples were too prominent. And Fisher, it was suggested, gave insufficient credit to previous statisticians, or warning to his readers that certain of his themes were still contentious.[59] Such mathematical niceties, however, were not at issue for social scientists, who saw Fisher's book as a route to objectivity and thus legitimacy. Following a favorable review by Harold Hotelling in America, *Statistical Methods for Research Workers* became a popular and influential textbook.[60] Within a couple of years, increasing numbers of statisticians were making their way to Fisher to learn statistical methods, and by the end of the 1920s Rothamsted was a rival to University College. This cemented Fisher's already secure position. He was given plenty of opportunity to pursue his own interests, and encouraged in his domestic

---

[56] Fisher 1925b, p. 2.
[57] See chapter 2, footnote 20; also Gigerenzer 1989, chapter 3.
[58] Fisher 1925b, p. 3.
[59] Box 1978, p. 130.
[60] Hotelling 1927. The book was widely translated and had sold nearly 20,000 copies in eleven editions by 1950. See Yates 1951.

mice- and poultry-breeding experiments.[61] In 1929, he was elected Fellow of the Royal Society.

3.8. THE DEVELOPMENT OF FIDUCIAL PROBABILITY

Fisher continued to work on estimation during the late 1920s, strengthening and widening the method of maximum likelihood. In *Statistical Methods for Research Workers*, he had again distinguished likelihood and probability as measures of uncertainty, and criticized first inverse probability (which was "founded upon an error, and must be wholly rejected"), and second the long history of attempts to "extend the domain of the idea of probability to the deduction of inferences respecting populations from assumptions (or observations) respecting samples."[62] Yet the doctrine of inverse probability was far from moribund. Statisticians, however sheepishly, were still using it to address specific problems. And although Chrystal earned Fisher's praise in cutting the method from his algebra textbook, it was widely included in others. Fisher attributed this persistence to the natural, though incorrect, assumption that since a scientific law is arrived at through a process of induction, so its reliability must be expressible in terms of probability. The popularity of his own method of maximum likelihood was also, somewhat ironically, to blame. Although clearly explained as an objective estimation procedure in *Statistical Methods for Research Workers*, the method was habitually taken to be nothing more than an application of inverse probability with uniform priors, and was widely used as such by geneticists for linkage estimation.[63] Fisher had shown that probability – defined in terms of frequencies in infinite populations – was inapplicable to hypotheses, and had pointed out that many inferences concerning parameters, such as linkage values in genetic problems, were simply inexpressible in terms of probabilities. Yet inverse probability and the associated epistemic concept of probability were still seductive. They promised an inferential method that was more direct and coherent than Fisher's weaker replacement of likelihood and the significance

---

[61] Box 1978, p. 97. Unanticipated problems with crop experiments at Rothamsted were addressed in a Christmas revue song: "Why! Fisher can always allow for it, All formulae bend to his will. He'll turn to his staff and say 'Now for it! Put the whole blinking lot through the mill."

[62] Fisher 1925b, p. 10.

[63] Neyman was one notable statistician who thought maximum likelihood a 'disguised appeal' to inverse probability (E.S. Pearson 1966). See also Yates and Mather 1963; Edwards 1972. Fisher credited this 'accidental association' as "perhaps the sole remaining reason why [inverse probability] is still treated with respect" (Fisher 1930a, p. 532).

test. This, up to the publication of *Statistical Methods for Research Workers*, was additionally in too mathematical a form to be widely understood. Even in the 1922 paper, Fisher's emphasis on large samples and data reduction obscured the process of inference from samples to populations.[64] Philosophers especially, doubtless encouraged by Pearson's public pronouncements, took statistical procedures, from regression analysis to significance tests, as merely methods of summarizing data. Such drudge work was commendable, but far from their loftier efforts at logical inference.[65] *Faute de mieux*, inverse probability survived. It was clearly not enough merely to destroy the inverse edifice, something at least as substantial had to be erected in its place.

Notwithstanding Fisher's declaration of 1922, it turned out that some sorts of frequency probability statements *could* be attached to hypotheses. Fisher drew on an observation the plant physiologist E.J. Maskell had made at Rothamsted in the early 1920s. The mean and modal values of a parameter's distribution are not in general equal to the mean and modal values of a function of that parameter – the mean squared value, for example, is not the same as the square of the mean value – but the median, or indeed any value that divides the distribution into specified fractions, will survive the transformation. This meant that the error limits of a population's mean could be expressed as the corresponding t-distribution percentiles rather than in terms of the standard error limits.[66]

In symbols, consider a sample of n observations, x, drawn from a normal population of mean μ. The mean value of the observations, $\overline{x}$, and the sample standard deviation, s, are defined as

$$\overline{x} = \frac{1}{n} \sum x$$

and

$$s^2 = \frac{1}{n-1} \sum (x - \overline{x})^2$$

[64] Indeed, Fisher had written in that paper that "briefly, and in its most concrete form, the object of statistical methods is the reduction of data."

[65] In his conspectus of probability, Keynes barely mentioned the statistical work of Galton and Pearson. He described statisticians as confining themselves wholly to the statement of correlations in observed data rather the inference from these to the correlations in unobserved data. Broad thought that in these remarks Keynes had "exactly hit the nail," and that Keynes was breaking new ground in his attempt to build an inductive theory of statistical inference. See Keynes 1921; Broad 1922; also Conniffe 1992.

[66] See Box 1978, p. 254; Zabell 1992, footnote 6, p. 371. Student made a similar observation in his original paper of 1908. On fiducial probability, see also Edwards 1976.

Student had shown that the distribution across different samples of the statistic t, where

$$t = \frac{(\bar{x} - \mu)\sqrt{n}}{s}$$

depends only on n, the size of the sample. Thus for any value of n, a value of t can be calculated such that this value will be exceeded in p percent of cases. Now, because t is a continuous function of the mean, μ, the conditions

$$t_1 < t < t_2$$

and

$$\bar{x} - \frac{s t_2}{\sqrt{n}} < \mu < \bar{x} - \frac{s t_1}{\sqrt{n}}$$

are equivalent, and must therefore be satisfied with the same probability. The error on the population mean can then be expressed by substituting the values of t for any chosen level of p. For $n = 10$, t has a 5% chance of falling outside the limits ± 2.262, so the 5% limits on the mean are

$$\mu = \bar{x} \pm 0.715s$$

In the same way, the probability that μ lies between any specified values – its probability distribution, in other words – can be calculated given the particular sample observed. This distribution, as Fisher later put it, "is independent of all prior knowledge of the distribution of μ, and is true of the aggregate of all samples without selection."[67]

As it stood, this was not worth publishing. But by 1930, Fisher realized the idea could be generalized. Consider a statistic T, whose random sampling distribution is "expressible solely in terms of a single parameter" θ, the value of which could have been estimated from the observations using the method of maximum likelihood. "If T is a statistic of continuous variation, and P the probability that T should be less than any specified value, we have then a relation of the form

$$P = F(T, \theta)$$

If we now give to P any particular value such as .95, we have a relationship between the statistic T and the parameter θ, such that T is the 95% value corresponding to a given θ, and this relationship implies the perfectly objective fact that in 5% of samples, T will exceed the 95% value corresponding to

[67] Fisher 1935c, p. 392.

the actual value of θ in the population from which it is drawn. To any value of T there will moreover be usually a particular value of θ to which it bears this relationship; we may call this the 'fiducial 5 per cent. value of θ' corresponding to a given T."[68]

To put it another way, assume the method of maximum likelihood has been used to estimate the value of some continuous single-parameter statistic. One could produce a probability function relating this statistic to particular values of the parameter. That is, a probability that some statistic should be less than a given value could be produced in terms of the statistic and the corresponding parameter of interest. At each value of the probability there is a definite relationship between the statistic and the parameter, reflecting that the statistic corresponds to that parameter with the chosen level of probability.

Fisher unveiled fiducial probability in 1930. As a weapon against inverse probability it was ideal. Fiducial probability could be used for inference, yet had a clear frequency interpretation.[69] (At the 5% level, the true value of the parameter will be less than the fiducial value corresponding to the observed estimate of the statistic in "exactly 5 trials out of 100."[70]) Fisher started his exposition with what he hoped would be a final blast against inverse probability. "I know of only one case in mathematics of a doctrine which has been accepted and developed by the most eminent men of their time, and is now perhaps accepted by men now living, which at the same time has appeared to a succession of sound writers to be fundamentally false and devoid of foundation. Yet that is quite exactly the position in respect of inverse probability."[71] Even a frequentist prior was mistaken: it assumes

---

[68] Fisher 1930a, p. 533. A valid statement of probability must be based on all available information, so T must be sufficient in θ for the fiducial argument to apply. Also, Fisher's notation does not distinguish between the statistic T and its estimated value t. Zabell's 1992, p. 372, gloss is more precise: "if $F(t, \theta) =: P_\theta[T \leq t]$, and if for each $p \in [0, 1]$, the relation $F(t, \theta) = p$ implicitly defines functions $\theta_p(t)$ and $t_p(\theta)$ such that (i) $F(t_p(\theta), \theta) = p$ and (ii) $\theta_p(t) \leq \theta \Leftrightarrow t \leq t_p(\theta)$, then $P_\theta[\theta_p(T) \leq \theta] = p$."

[69] "I imagine that this type of argument, which supplies definite information as to the probability of causes, has been overlooked by the earlier writers on probability, because it is only applicable to statistics of continuous distribution, and not to the cases in regard to which the abstract arguments of probability theory were generally developed, in which the objects of observation were classified and counted rather than measured, and in which therefore all statistics have discontinuous distributions." Fisher 1930a, p. 534.

[70] In fact, the frequency interpretation of fiducial probability turned out to be problematic, and Fisher later softened this sort of explicit statement. It is not strictly true, for example, in the example given earlier for the mean of the sample drawn from a normal distribution of unknown standard deviation. See Yates and Mather 1963, p. 103.

[71] Fisher 1930a, p. 528. Significantly, this 1930 paper introducing fiducial probability was titled "Inverse Probability."

78

"that cases are equally probable (to us) unless we have reason to think the contrary, and so reduces all probability to a subjective judgment."[72] Fisher then outlined the new probability before pointing out its logical distinction from a statement of inverse probability. In addition to depending on the prior probability, a posterior distribution for the population statistic will hold only for the statistic derived from a specific set of sample values. In contrast, the fiducial procedure is relative to this variability, and gives a value (for that fraction of cases that the true value of our parameter would be less than that estimated from a sample set) regardless of what the sample set in fact is. This is not only more useful and general than an inverse probability statement, declared Fisher, but an objective fact about the parameter, since it is independent of any assumption about prior distributions or specific sets of sample values.[73]

3.9. FISHER'S POSITION IN 1932

From his earliest publications, Fisher deplored the inconsistency of statistical practice and called for reform of the conceptual foundations of his subject. Though aware of the shortcomings of inverse probability, it was not until provoked by Pearson that he scrutinized, and dismissed, not only the Principle of Insufficient Reason, but inverse probability and the associated epistemic interpretation more generally. The model of statistics with which he proposed to replace it was suggested by Mendelian genetics, and reinforced by his experiences in agricultural science. Probability was a feature of nature, a precise frequency in a population of well-behaved and independent measures. Such a frequency was inapplicable to hypotheses or theories. Fisher initially promoted the weaker combination of likelihood and significance test as an inferential method to replace inverse probability. In 1930, however, he triumphantly announced fiducial probability as a powerful alternative.[74] Clearly a frequency, fiducial probability could be used to attach probability values to hypotheses without the introduction of prior probabilities.

---

[72] Fisher 1930a, p. 528.

[73] Fisher noted that tables of values of the variates at any given fiducial percentage could be constructed, and gave one for the correlation coefficient. For any sample correlation coefficient, the 5% fiducial value of the population correlation coefficient can easily be read off. Although tables for the mean or standard deviation fiducial intervals might have seemed more obvious illustrations, the correlation coefficient was the bone of contention in the cooperative study episode.

[74] Zabell describes fiducial probability as the "jewel in the crown" of Fisher's inference work. Fisher still regarded it as superior to the "weaker" combination of likelihood and significance test in 1956. See Zabell 1992.

Accompanying this constructive work were Fisher's efforts to discredit inverse methods. He was a master of polemic. He was also adroit in the selective mobilization of history to bolster his new inferential techniques. In 1922 he had presented himself as a lone voice criticizing inverse probability against its habitual use by eminent statisticians such as Edgeworth and Pearson. Later, when more confident of his rival methods, he declared either that the doctrine had faded gradually towards the end of the nineteenth century, or had collapsed abruptly – "universally abandoned" – after the irrefutable criticisms of Venn, Boole, and Chrystal.[75] That this was all a 'literary technology' has been argued persuasively by Zabell.[76] Fisher's grasp of this history of his field was detailed. Yet he elided the subtleties and nuances of the trio's objections to inverse probability in order to strengthen their rhetorical support. Boole did not object to empirical Bayes methods or the logical interpretation of probability; Venn reserved his scorn for the Rule of Succession, which Fisher tended to support; Chrystal criticized Bayes's theorem rather than the Principle of Insufficient Reason itself. Despite Fisher's attacks, then, inverse probability was weak, but still viable, in 1930.

[75] Chrystal's presence in the trio owed more to his didactic potential and personal relevance for Fisher than his influence on the history of the subject: Chrystal's *Algebra* was the chief text for the first part of the Cambridge Mathematical Tripos.

[76] See Zabell 1989b and the commentaries Plackett 1989 and Barnard 1989. (See also Aldrich 1995, p. 48, for an alternative view, in which Fisher's version of the history of the subject is rather an expression of his own learning process: "Yet Fisher is laying his ghosts. He projects his own confusions onto the world, and then clears them away.") For more on 'literary technologies' in science, see Bazerman 1988.

4

Harold Jeffreys and Inverse Probability

4.1. JEFFREYS'S BACKGROUND AND EARLY CAREER

Harold Jeffreys's scientific interests were well developed by his arrival in 1910 at St. John's College Cambridge on a mathematics scholarship. Some interests, such as a lifelong fascination with celestial mechanics, had been inspired from the wide reading of an only child, which included popular scientific works by Sir Robert Ball, G.F. Chambers, and Sir George Darwin. Others were influenced by the elemental countryside of his Northumberland home. Greasy slag heaps and mountains of ore dominated the landscape of the colliery village of Fatfield, and the air was thickened with the smoke from nearby iron and shipbuilding works. The area was rich in both geological and botanical terms. Rocks from any period from the Silurian to the Jurassic were visible within a day's train journey, while certain plant species were unique to the coal measures. Encouraged by his parents, both able gardeners, Jeffreys took up naturalism early, recording in small black notebooks wildlife and rock formations studied during cycling trips around the locality.

Jeffreys was unusually well prepared for undergraduate study at Cambridge, having already obtained a B.Sc. in the general science course from Armstrong College (then part of Durham University, now of the University of Newcastle). The catholic approach at Armstrong had fostered Jeffreys's early interests: the geology course was supplemented by regular excursions to the local countryside, and Jeffreys was encouraged to apply his year of chemistry to study the development processes of his hobby of photography.[1] Nevertheless, the high standard of the mathematics at Cambridge came as a shock. Financially, things were difficult too. Jeffreys's schoolteacher parents were not particularly well off – the family home was perpetually damp and unsanitary – and although the scholarship at St. John's was supplemented by discretionary money from Durham County Council and Armstrong College,

---

[1] Jeffreys's first two published papers, concerning a sulfite toning process that he considered patenting, appeared in the *British Journal of Photography* in 1910 and 1911.

81

his circumstances were straitened. Scholarship income was linked to academic performance, however, and success in the third year exams brought a measure of relief. With Wrangler status following Part II of the Mathematics Tripos in 1913 came various prizes, and, importantly, an extension of the college scholarship making possible a fourth year of study.

With the degree of Ph.D. not offered until 1921, Cambridge awarded no qualifications between the B.A. and the Sc.D.[2] Instead, fourth-year original research was a stepping-stone to professional recognition. Jeffreys had not neglected his early interests: encouraged by the young botanist E.P. Farrow, who often accompanied him on his north country cycling trips, he published a number of works of a conservationist tone on vegetation and soil types in his native Durham while a student at Cambridge, and served as a referee to a new and still current natural history magazine, *The Vasculum*, founded in 1915 by his friend the botanist and entomologist J.W.H. Harrison. But following the lectures of Ebenezer Cunningham and Arthur Eddington – the chief British exponents of Einstein's relativity theory at the time – and discussions with the physicists Newall and Larmor, Jeffreys committed himself firmly to a study of the planets.

The move from the mathematics tripos to physics was common at Cambridge, especially later during the first years of quantum mechanics, and resulted in a distinctly analytical approach to theoretical problems. Jeffreys was particularly creative. From classical mechanics he would construct ingenious mathematical models of his physical systems, then apply powerful and elegant approximation techniques to yield tractable solutions. (The statistician Dennis Lindley has justly described Jeffreys as a "master at the gentle art of approximation," a necessary skill in pre-computer days.[3]) The Earth and Moon could each be considered a system of homogeneous shells surrounding a core, for example, and thus the concepts of friction, viscosity, and vibration linked to astronomical observables such as excess ellipticity and variation of latitude. The combination of effects was complex, and with free parameters often poorly known the agreement with empirical data variable. Yet this early attempt to build a systematic mathematical model of the structure and dynamics of the planets and solar system from a compound of

---

[2] In the 'dossier' of personal notes Jeffreys complied for the Royal Society, he described the award of Sc.D. as "more or less equivalent to being proposed for the Royal Society." The M.A. degree was available, but was as now more a formality than a matter of academic achievement.

[3] Lindley 1986. Each calculation in Jeffreys's 1932 paper on the rejection of observations took him six hours with a calculating machine. Jeffreys 1932a, p. 85.

seemingly prosaic mechanisms was to become a Jeffreys characteristic. His first work – rewarded by a college prize in 1912 – treated the precession and nutation of the Earth.[4] By gradually incorporating fluid and solid effects, such as the damping effect of the tides, he was able to account for a wide variety of astronomical and meteorological observations, and thus to speculate on the origins of solar system.[5]

4.2. THE METEOROLOGICAL OFFICE

Jeffreys's early efforts were sufficient for election to a college fellowship in November of 1914. Yet a fellowship did not guarantee a teaching post, and from 1915 to early 1917 Jeffreys worked with little success on war-related problems at the Cavendish laboratory. His specialist abilities were not entirely irrelevant to the war effort. Since the success of a maritime engagement depends in part on sea conditions, the Royal Navy had asked Sir Napier Shaw, then director of the Meteorological Office, to investigate the prevailing weather and tides of the North Sea theatre. His own research specialty being in ventilation and air currents, Shaw approached his friend Newall for names of likely candidates for a theoretical approach. Newall suggested Jeffreys.[6]

Hastily boning up on meteorology and fluid dynamics, Jeffreys set off to South Kensington in early 1917 as Shaw's personal assistant. The subject was perfect for him. Hydrodynamics appeared only briefly in the Cambridge course, and was not yet a mathematically mature field with distinct theory and methods. Jeffreys was able to develop his own models to address the effect of tides and wind on the sea conditions at the French coast. One or two other specific problems filtered through from the front – calculating trajectories in gunnery, for example – but generally the work load was slight, and Shaw allowed Jeffreys to spend most of his time developing general models for hydrodynamic systems. Jeffreys shared an office with G.I. Taylor, a bright and inquiring young Cambridge physicist who two years earlier had published

---

[4] The axis of the Earth is not perpendicular to its plane of rotation about the sun, but is inclined at an angle of 23.5 degrees. Due to gravitational attractions of the sun and moon on the Earth's equatorial bulge, the axis slowly rotates to sweep out a cone every 26,000 years. Imposed on this 'precession' is a slight nodding motion of the axis called the nutation. The nutation is mainly due to the pull of the moon, and has a principal period of around 19 years.

[5] For Jeffreys's early work see Cook 1990; B. Jeffreys 1991, 1992; HJ (in particular the 'dossier' of personal notes Jeffreys complied for the Royal Society).

[6] HJ-MM.

seminal work on eddy motion in the atmosphere, based on kite-flying experiments in Newfoundland. Gases are fluids too, and Jeffreys discovered that Taylor's work could be turned to the motion of the sea. By incorporating the effects of eddy viscosity and convection, Jeffreys could construct a mathematical model of wind-driven wave creation in satisfactory agreement with observations of the North Sea.

The period at the Met Office did not blunt Jeffreys's interest in astrophysics. Instead, his theory of tides provided new ways to study the Earth and the planets. Eclipse records indicated a relative non-periodic, or secular acceleration of the Moon, in part attributable to the slowing of the Earth. Jeffreys extended Taylor's 1919 study of turbulence in the Irish Sea to show that the worldwide effect of tidal friction on the beds of shallow bodies of water, such as the Bering Sea, was sufficient to account for the effect.[7] His work also allowed a new estimate of the age of the Earth. This issue was still lively fifty years after William Thomson had argued from energetics against Darwin's theory of evolution. Thomson's industrial background, quite apart from his moral distaste at the implications of Darwin's theory, had led him to a thermodynamic conception of planetary change incompatible with the vast periods of uniformity necessary for natural selection.[8] However, although Thomson's work became one of a number of scientific and religious objections to Darwin's theory that led to its displacement by the turn of the century in favor of neo-Lamarckism and other fast mechanisms of directed variation, his arguments had largely been defused by the time Jeffreys worked on the problem. Physicists such as Soddy and Rutherford had been quick to see the significance of Pierre Curie's announcement in 1903 that heat was produced by the radioactive decay of radium. Even a minute concentration of radium could generate enough warmth to balance the cooling effect of energy radiated from the Earth's surface. With new data on the composition of radioactive elements in different types of rocks and minerals, physicists began around the time of Thomson's death in 1907 to calculate accurate values for the age of the Earth. Arthur Holmes, based within walking distance of Jeffreys at Imperial College, had in 1911 produced a value of around 1.5 thousand million years. Following discussions with Holmes, Jeffreys was able to produce a consistent value by combining his work on the slowing of the Earth with a tidal theory of the solar system, that the planets had formed by condensation from a filament of gas drawn from the sun by a passing body. This was a theory

---

[7] Jeffreys 1920.
[8] See Smith and Wise 1989, pp. 579–611.

that he, and James Jeans independently, had begun to propose from around 1917.[9]

4.3. DOROTHY WRINCH

The exposure to a wide range of scientists engaged on real research problems in cosmopolitan London started Jeffreys pondering the institution of science in Britain. Provoked by a popular debate on the role of research in wartime, he involved himself from the summer of 1917 with a small group of scientists, mostly from Cambridge and London, who were aiming to form a trade union to protect the career interests of scientists and technicians. At the same time, he began to question the fundamentals of the scientific world view. Science was generally supposed to deal in certainties, but how much confidence could be placed in his mathematical models of the solar system, in which the same equations that governed the viscosity of water in a bucket were extrapolated to cover the condensation of the planets? And what of Einstein's recently-proposed rival to Newton's law of gravitation? How was a physical law of that sort evaluated?

Also involved in the fledgling National Union of Scientific Workers was Dorothy Wrinch, a young lecturer at University College London. Though she taught mathematics, in which she had graduated as the sole Wrangler from Girton in 1916, Wrinch had become fascinated with epistemology after attending Bertrand Russell's lectures in 1913. Indeed, she had taken Part II of the Moral Sciences Tripos in order to remain at Cambridge to study symbolic logic with Russell, and had been one of a small group that met with him after 1916 for informal discussions on the foundations of mathematics. After Russell had helped in obtaining scholarship money and the occasional paid translation work, Wrinch became a sort of personal secretary, checking references and organizing his academic duties. During Russell's 1918 spell in Brixton prison for activities relating to his pacifism, she supplied him with books and journals, dealt with his correspondence, and kept him abreast of the latest philosophical meetings.[10]

Confident and able women were unusual in science, and with her connections to Russell and his permissive and elite social and intellectual circles, Wrinch must have seemed an exotic figure to the diffident and retiring Jeffreys. Yet they found common interests. Both were attracted by the power

---

[9] Holmes 1913, Jeffreys 1921c, also Jeffreys 1925, pp. 38–9.
[10] Abir-Am 1993, 1987, pp. 243–8; Monk 1996, p. 525.

and generality of set theory and transinfinite series. Both too had developed a curiosity toward the new study of psychoanalysis, Wrinch following Russell, with whom she had corresponded on the subject, and Jeffreys probably influenced by his botanist friend, E.P. Farrow.[11] The two began to discuss the status of general laws.

As a theorist, Jeffreys worked largely alone, and had had little exposure to the instrumentalist parole that prevailed at Cambridge. Instead, his philosophy of science was heavily influenced by Karl Pearson's brand of positivism. It was Farrow who in 1914 had introduced Jeffreys to Pearson's *The Grammar of Science*, and the two had discussed it on their cycling tours. For Pearson, a scientific law was a shorthand expression of the correlation between certain groups of perceptions. The more fundamental the law, the more economical this description. Laws were not explanatory, nor, since there was no necessity for past events to recur in the future, were they prescriptive. Laws based on past experience could be used to predict future events, but such predictions were at best defeasible, in the nature of beliefs, or guides to action, rather than knowledge.[12]

This was heady stuff. Yet perhaps there was indeed nothing special about scientific laws: psychologists taught that all learning was merely an individual's attempt to make sense of experience. And the turmoil in the physical sciences seemed to support Pearson's contention that even apparently bedrock knowledge was only tentative. Twenty years previously, the luminiferous aether had been taken as an objective reality; now it seemed that Newton's laws of motion might be similarly revealed as nothing more than a construction of the mind. Nevertheless, some aspects of Pearson's philosophy worried Jeffreys. The success of his solar system model had taught him that the extrapolation of scientific laws, though indeed never a matter of certainty, was something more than guesswork. Data could be used to infer causes. Jeffreys

---

[11] Farrow was to publish a number of papers on the subject from the mid-1920s, followed in 1942 by a popular book, *A Practical Method of Self-Analysis*. For more on Jeffreys and psychoanalysis, see Cameron and Forrester 2000, who cover his interest and involvement in the field as part of their rewarding study of the loose psychoanalytic network that formed around the botanist A.G. Tansley, an early patient of Freud and an enthusiastic promoter of his ideas in the UK. The network was primarily made up of Cambridge-based scientists who shared a view of "the new ideas and practices associated with psychoanalysis as a natural extension of the 'scientific attitude' of careful and empirical inquiry into the nature of human beings." It included Farrow, who had been a student of Tansley's in Cambridge before World War I.

[12] Even at the end of his life, Jeffreys regarded the *Grammar* as "the best account of the scientific method," the "connecting thread" between all his work. See, for example, Jeffreys 1974.

detected a tension between Pearson's phenomenalism and his concession that descriptive laws might, even only probabilistically, lead to reliable predictions. Pearson had argued that it was experience of the persistence of law-like relations in general that justified the assumption that a law-like relation will also persist in any specific circumstance. But surely this was circular? A belief in such stability must itself presumably have been inferred from experience. How then could general laws ever become established? Jeffreys discovered that Wrinch too had been troubled by such questions while studying induction and probability in W.E. Johnson's lectures on advanced logic at Cambridge.[13] And Russell had taught her that mathematics could not provide an ultimate basis for physical theories. Perhaps she and Jeffreys could work together to find a set of principles to justify a general theory of inference from experience?

4.4. BROAD'S 1918 PAPER

The stimulus was a paper by the young philosopher C.D. Broad, who had also started to consider the relation between induction and probability after attending W.E. Johnson's lectures.[14] The syllogism is apparently a powerful tool. All men have two legs; Archimedes is a man; Archimedes has two legs. Yet it hangs on the generality of the major premise. Indeed, there is no new knowledge here: the major premise already contains the conclusion. Such generalizations are necessarily conjured from a limited number of specific observations. Thus the seemingly ineluctable deduction of testable predictions from general scientific laws conceals the formally illicit process of induction.

Broad wondered whether the laws of probability could resolve the situation and justify our confidence in scientific induction. He turned to the standard balls-in-urn model. Laplace had assumed in his treatment of induction that the urn was infinitely large. But if the model was to apply to general laws of the type 'all crows are black,' this needed modification. Since there are only a finite number of crows in the world, Broad addressed the case of a finite number, n, of balls that, once drawn from the urn, were not replaced. He produced the same expression Laplace derived for the infinite case: the probability of the next crow's being black after the successive observation of m black crows is $(m + 1)/(m + 2)$ whatever the total population of crows.

---

[13] Jeffreys 1931, footnote on p. 177.

[14] Broad 1918. Broad, who went on to develop an interest in the paranormal, was later described by A.J. Ayer as one of the two important Cambridge philosophers of the century (the other being J.E. McTaggart). Ayer 1992, p. 48.

But what of the general law? What is the probability that all n crows will be black? Broad stated the answer to be $(m + 1)/(n + 2)$, and pointed out that in consequence the probability of the general law will never become appreciable, even if every crow seen is black, until we observe a fair class of the total population of crows. If drawing balls from a urn is analogous to the process of scientific induction, we could never adduce a reasonable probability for any scientific law.[15]

The problem of the general law had not arisen in Laplace's infinite treatment. Broad's finite case, in contrast, allowed the probability of a general law to be calculated and thus provided a more promising analogy to scientific induction. Well, but is induction by enumeration – the balls-and-urn model – anything like the process by which knowledge is inferred? Despite the historical debate over this question, and though drawing his logical interpretation and relational treatment of propositional probabilities directly from the skeptical Boole, Broad was satisfied both with the quantification of probability and with enumeration as a typical rather than a special case of general induction. How then to explain the muddle over general laws? Broad's conclusions were pessimistic. The concept of probability was a necessary component in a formal justification of induction: it applies to uncertain relationships, can account for occasional inductive mistakes, and accords with our experience about the nature of learning. Yet the Rule of Succession flowed inexorably from the calculus of logical probability. The concept, therefore, was insufficient to account for our degrees of belief. Some factor additional to the principles of probability was needed if a statement of induction were to be formally valid. Such a factor must be a function of the universe itself, and thus could only be established by induction. We are where we started.[16]

---

[15] The assumption that each crow has the same probability of selection is dubious: only crows in the immediate vicinity of the observer are practicable selections. Further, the calculation neglects the possibility that certain crows will be unwittingly observed several times. Broad showed, however, that the assumptions of a more realistic model – drawing with replacement from a restricted part of a vast urn – unfortunately act to constrain the probability further and make general laws still less likely.

[16] Broad regarded the application of probabilities to propositions on data as a sophisticated generalization of the balls-and-urn sampling model, and inverse probability as "the favorite instrument of all advanced sciences," its use borne out by the history of science. Yet his mathematical result was far from original. The coincidence of the finite and infinite sampling cases had been derived by the Swiss mathematicians Pierre Prevost and Simon L'Huilier in 1799, rediscovered independently several times during the nineteenth century, and even remarked upon by Todhunter in his well-known *History of the Mathematical Theory of Probability*. Nor had the problematic relevance of this sort of enumerative induction for general laws been overlooked. Fries cited it in 1842 as a convincing reason

Sampling clearly had to be revisited. Though Jeffreys's experience of probability at Cambridge had been restricted to a course on the theory of errors taught by the astronomer Arthur Eddington, he had learnt inverse probability as a schoolboy from his father's copy of Todhunter's *Algebra*, and like the young Fisher had found no reason to doubt it. Yet casting around for ideas, Jeffreys was amazed to learn that this form of probability – used uncritically and "absolutely clearly" by Pearson – was apparently out of date. Mathematicians instead seemed to follow Venn in defining probability as a limiting ratio in an infinite sequence of events. Jeffreys was horrified. From his work with series expansions in physical problems, he knew that infinite series provided only approximate solutions; they were no basis for a definition. Further, such a definition was not just practically but mathematically untenable, since limiting ratios could not be guaranteed for infinite sequences.[17]

Wrinch and Jeffreys's chief concern, however, was that the frequency approach was unnecessarily restrictive. Following Pearson, they took a probability to be a conditional relation between a proposition and data, a relation that reflected a 'quantity of belief' or 'quantity of knowledge,' and could be used directly to assess the credibility of empirical generalizations or scientific hypotheses. The attempt to ground probabilities in frequencies did not conform to these commonsense and customary applications of the concept of probability. How could the frequency approach confirm laws? With probability restricted to series "one could attach no meaning to a statement that it is probable that the solar system was formed by the disruptive approach of a star larger than the sun, or that it is improbable that the stellar universe is symmetrical, for the indefinite repetition of entities of such large dimensions is utterly fantastic."[18]

---

why induction could not merely be a matter of probability, and Keynes referred to the calculation of the "ingenious German, Professor Bobeck" that even given 5,000 years' worth of consecutive repetitions, the probability of the sun's continuing to rise in unbroken sequence for another four thousand years was, according to Laplace's analysis, no greater than $\frac{2}{3}$. (Actually nearer $\frac{1}{2}$, notes Zabell.) Nevertheless, the result seemed not to be generally known and was taken by Broad at least to raise a severe doubt over the possibility of treating induction probabilistically (Broad 1918, pp. 402–3. See Todhunter 1865; Zabell 1989a, pp. 286, 307; Keynes 1921, p. 383; and Dale 1991 for previous discoveries of the coincidence between the infinite and finite cases. Zabell explains this coincidence with the concept of 'exchangeability').

[17] HJ-GAB, HJ (notes for interview with Dennis Lindley, 24/8/83), Wrinch and Jeffreys 1919, pp. 716–7.

[18] Wrinch and Jeffreys 1919, p. 716.

A logical justification of probability was not required. Better to take the epistemic notion as primitive, needing no definition.[19] What *was* needed for probabilistic induction was a numerical account of the way probabilities should be assigned to combinations of propositions and data. Wrinch knew the correct form of such a theory: simply the axiomatic style of Russell and Whitehead. Starting from the assumption that a positive real number could be assigned as the probability of each relevant combination of proposition and data, she and Jeffreys produced a set of axioms concerning the comparison of such probability numbers, and proceeded to derive expressions for their combination. These included Bayes's theorem, and the Principle of Insufficient Reason as a special-case corollary.[20]

Probabilistic inference could best be applied to balls-in-urn sampling. What then of Broad's problem that even favorable enumeration would never allow general laws of the form 'all crows are black' to attain a reasonable probability until the sample size approached the totality of cases? Wrinch and Jeffreys replied that the assumption made by Laplace and Broad, of complete ignorance of the urn's contents, is overly conservative for scientific inquiry; indeed, that this starting point, of setting equally-probable each initially possible permutation of balls in the urn, expressed from the outset a sharp prejudice against the prospects of the general law that *all* the balls were one color or the other. In practice, they noted, situations in which we can assume a uniform prior probability for the ratio of balls are rare. In everyday life we observe "a strong tendency for similar individuals to be associated."[21] Like things usually

---

[19] Wrinch and Jeffreys 1919, pp. 722, 725, 731. Probability is "an entity known to exist independently of definition, intelligible without definition, and perhaps indefinable." To retain the widest application and the commonsense meaning of probability "seems desirable at any cost."

[20] The axioms were that for each combination of proposition and data there is a single probability number; that this number is greater for more probable combinations; that the probability of mutually exclusive propositions on the same data is the sum of the individual probabilities; and that the largest and smallest values of probability correspond respectively to true and untrue propositions. (Wrinch and Jeffreys 1919, p. 720.) Wrinch and Jeffreys noted that the frequency theory was usually commended over any epistemic interpretation of probability for being based on fewer initial assumptions, but pointed out that when hidden assumptions about limits, independence, and so on were made explicit, the frequency theory was more axiomatically top-heavy than their own. (Wrinch and Jeffreys 1919, p. 729.)

[21] Wrinch and Jeffreys 1919, p. 726. See too Jeffreys 1931, p. 31. This observation was hardly novel. Compare with Price, who had noted 150 years earlier in his appendix to Bayes's essay that the regularity of nature "will cause one or a few experiments often to produce a much stronger expectation of success in further experiments than would otherwise been reasonable; just as the frequent observation that things of a sort are disposed together in

come grouped together: a bag of balls will more probably be all for cricket, or hockey, or tennis than a random mixture. Wrinch and Jeffreys conceded that to reflect accurately this tendency with a prior probability was difficult, but showed mathematically that with a large enough sample the precise choice of prior probability had little influence on the calculated posterior probabilities, and that large samples would usually be fair even if only a small fraction of the whole class. These results meant that in Broad's example it would be valid to infer that the fraction of non-black crows is small, and thus that the next crow to be seen will be black with high probability. This is all that is ever inferred in practice. Of course, this still leaves Broad's main point that the probability of the general law will remain low, because the number of observed instances in any case will always be vanishingly small compared to the total possible number. But Wrinch and Jeffreys pointed out that general laws of the 'all crows are black' sort *are* unlikely to be true in every instance. After all, exceptions or 'sports' – a term used by biologists to refer to rare deviations from type – always exist. Broad's law in fact speaks of certainty rather than probability, and could therefore only be arrived at by deduction from some non-probabilistic data, such as a 'principle of uniformity,' or Broad's *a priori* belief that certain properties are shared by all members of a class.[22]

The 1919 paper of Jeffreys and Wrinch is more a reactionary attack on the superficial attractions of the precise frequency definition than a precursor of modern Bayesianism. Yet in defending the epistemic interpretation of probability for uncertain inference, and in championing an axiomatic exposition, Wrinch and Jeffreys took an important forward step. Broad, like Poincaré and Jevons, had taken the Principle of Insufficient Reason as the self-evident truth required to ground a probabilistic approach; in consequence he was forced to reject the probability calculus as the sole source of our inductive belief. Wrinch and Jeffreys, however, in starting from axioms concerning comparability rather than the combination of propositions, were able to decouple the Principle of Insufficient Reason from the rest of the probabilistic machinery, and so widen the range of probabilistic induction to the predominant situations in which equally probable sets of alternatives could not be constructed. This was consistent with their epistemic rather than purely logical interpretation: probability theory was primarily to account for the scientific treatment of new data, not the assignment of initial attitudes to propositions. It was a model

any place would lead us to conclude, upon discovering there any object of a particular sort, that there are laid up with it many others of the same sort."

[22] "Such [laws] cannot be derived by means of probability inference, for [they deal] only with certainties." Wrinch and Jeffreys 1919, p. 729.

of scientific *method*, which, as Pearson had shown, was more important than the content. The Principle of Insufficient Reason was thus properly seen as a guide to initial probabilities, not as part of the methodological apparatus.

Wrinch and Jeffreys's presentation departed from that of Broad in another respect too. Although they did not give concrete examples of how prior probabilities should be assigned, their decision to publish for a scientific audience in *The Philosophical Magazine*, rather than following Broad in the philosophical *Mind*, demonstrated their commitment to probability theory as a tool for practicing scientists to assess and evaluate new theories and hypotheses, rather than for philosophers concerned with justifying inductive reasoning. It was practical experience, and not philosophical first principles, that could justify the choice of prior probabilities and thus our confidence in general laws. Balls, urns, and black crows were still to the fore in the paper, but also in evidence were scientific examples, such as the forms of the principle of relativity proposed by Einstein and Silberstein, used to illustrate the competition between contradictory laws that could be decided by inductive arguments.

4.6. AFTER THE FIRST PAPER

4.6.1. General relativity

The status of general laws of gravitation and motion, which had originally piqued Jeffreys's interest in probabilistic inference, soon seemed nearer to resolution. In 1919, the Cambridge astronomer Arthur Eddington – quondam lecturer of Jeffreys – organized expeditions to the coast of West Africa and northern Brazil to photograph the total solar eclipse of May 29. Although both Newton's laws and the newer analysis of Einstein predicted that starlight passing near the sun should be deflected by gravity, they differed over the amount. Each effect was too tiny to be observable against the background radiation from the sun. But if this sunlight were blocked by an eclipse, the starlight-bending should be visible on photographic plates, and the amount of bending would determine whether Einstein or Newton was right. The announcement of the result, which seemed to support Einstein's theory and his counter-intuitive model of space-time, provoked immediate controversy. Comment started to fill the letters columns of *Nature*, and in 1921 an entire issue of the journal, containing articles by Einstein and Eddington, was set aside for the subject.[23]

---

[23] There is a vast literature on this experiment and the history of relativity theory more generally. The picture is considerably more confused than presented here. Eddington's results were not initially taken as unequivocal. Some questioned the quality of the data;

Though some commentators criticized Eddington's results, or constructed elaborate thought experiments involving mirrors and beams of light that would supposedly allow the relative velocity of the Earth and the aether to be calculated, most sought to celebrate the breakthrough. One common theme was that in uniting the dimensions of space and time, Einstein had lifted the demarcation between physics and metaphysics. Relishing the opportunity to indulge in philosophical speculation, a good fraction of the correspondents to *Nature* revisited ancient questions of whether space could exist without constitution by an extended body, or vice versa, or whether energy and space-time were ultimate entities, or perhaps products of the universe's geometry.[24]

Jeffreys and Wrinch were both regular contributors to the debate. Wrinch had discussed the philosophical implications of relativity with Russell, from whom she had heard preliminary results of Eddington's expedition; Jeffreys had previously corresponded with Eddington on astronomical matters, and had rushed immediately into print following the announcement with his gloss that Einstein's theory was the only way to account for both the new results and the anomalous perihelion of Mercury.[25] The gaseous ramblings in the letters pages they considered meaningless. Instead, a dose of Pearson was in order. Science is about describing past experiences and predicting future ones. Entities are merely defeasible concepts that order sensations. There is no need to invoke untestable hypotheses: it is safer simply to assume the differential equations of electromagnetic waves, for example, than to create an elastic aether in order to justify them. Talk of 'space' is superfluous; the aether too is a "metaphysical concept of no utility."[26]

Wrinch and Jeffreys's approach was exemplified by a joint paper that appeared in the special issue of *Nature*. Despite obtaining crucial empirical evidence, Eddington had declared that Einstein's theory was really a sort of geometry, and could be deduced from *a priori* principles alone. Wrinch and Jeffreys countered that science is never a matter of introspection or necessity, but the inferential process of generalization from primary sense data. Dynamics – and thus the theory of relativity – is no different. Geometry is indeed the starting-point, but in its literal, Earth-measuring sense of mensuration, the "science of the relations between measurements of distance in rigid

other explained it away with alternative light-bending theories. Though the matter was finally settled in Einstein's favor in 1921, Brush has suggested that more persuasive than the light deflection results was the success of Einstein's theory in accounting for the well-known anomaly in the perihelion of Mercury (Brush 1989. See also Mayo 1996).

[24] See, e.g., Bonacina 1921 and Synge 1921, and Jeffreys's replies, Jeffreys 1921a and Jeffreys 1921b respectively.

[25] Jeffreys 1919.

[26] Jeffreys 1921a, p. 268.

bodies," rather than in the Euclidean sense of a purely logical progression from *a priori* principles.[27] Any theory of mensuration must be verifiable and practically applicable; Nature is quantified by the actions of scientists with measuring rods. Since the dimensionless points and infinitely extendible lines of Euclid's geometry are a generalization from experience, geometrical theories must follow, not precede, mensuration. 'Space' is therefore not a causal or explanatory concept. Einstein himself, noted Wrinch and Jeffreys, tended to treat space as a hypothetical construct, and one that had to face the empirical data like any other. His great contribution, they said, was to recognize that Euclid's generalization could not account for real experience, and to replace clocks and rigid bars as a basis for comparison with a new measuring standard based on light signals.

4.6.2. *The Oppau explosion*

Though participating in the relativity debate, Jeffreys and Wrinch were not neglecting the practice of science. Around 1919, Jeffreys had found a new target for his mathematical armory. In speculating on the origins of the solar system he had focused on effects that dissipate energy from the Earth–Moon system. What about earthquakes? Did the energy lost in earth tremors constitute a significant fraction of the total? As a quantitative scientific discipline, seismology was in its infancy. The extant theory – due to Rayleigh, Love, and subsequently Milne – was based on Poisson's 1828 demonstration that elastic and isotropic solids transmit shock waves with two distinct components. The faster of these oscillates longitudinally to the direction of propagation, like a sound wave, and the slower transversely, like a wave on a shaken rope. Extending this theory was not straightforward. The greatest obstacle was the lack of good numerical data on earth tremors. Required was a record of the motion of the ground relative to some fixed point as an earthquake passed. Milne's solution had been his invention, around 1910, of a new instrument, the seismograph. The device comprised a mass sensitively suspended by insulating springs – the 'fixed point' – coupled to a fine stylus that played against a rotating drum to produce a 'seismogram' of relative ground motion with time. The seismogram traces were complex and difficult to decipher. If the Earth were homogeneous, earthquake waves would radiate uniformly, traveling in straight lines direct from the source to the recording station. But the Earth is not homogeneous. Instead, it is split into roughly concentric regions – such as the core, crust, and mantle – and made up of different types of rock. The

---

[27] Wrinch and Jeffreys 1921b, p. 807.

interfaces between these regions can reflect or refract incident waves. Before reaching the recording station a wave might be reflected and refracted many times, or give rise to secondary or dispersive waves. The observed seismograph is a composite of all the received signals.

Even so, some inference should be possible. The longitudinal component, termed the primary or P-wave, and the transverse component, termed the secondary or S-wave, each produce characteristic seismogram traces.[28] The distance between the first-arriving of each kind can be measured and, knowing the speed of the revolving drum and the travel speed of the waves, converted into a rough estimate of the distance between the epicenter of the quake and the recording station. With seismograms from more than one research station, the time of the event and position of the epicenter can be triangulated accurately.

In theory, anyway. In practice, the early machines were unreliable, each with different operating characteristics. Some were more efficient and precise than others. Also, there were few seismological observatories. Yet the earthquake data were surprisingly useful. Once the time and location of the source had been established, the problem could be inverted to yield precise travel times of the shock waves. Since waves pass through different rock formations with different speeds, these travel times turn out to be a function of the geological structure of the Earth. From evidence of this sort, Oldham had detected the Earth's core in 1906 and Mohorovičić had inferred the existence of a thin, 30 km mantle shortly after.[29]

Jeffreys was excited at the possibilities for studying the Earth's interior, but frustrated by the quality of the data. Not only were earthquake travel times scarce, but their analysis haphazard. In 1918, H.H. Turner in Oxford had started to publish the *International Seismological Survey*, a quarterly tabulation of worldwide earthquake records, and at least estimated depths and times of earthquakes from the results; even so, his methods were crude. Clearly, better statistical procedures would increase the accuracy of the data and hence the reliability of inferences about the structure of the Earth. Jeffreys and Wrinch soon had an opportunity to try such an approach. Early in the morning of September 21, 1921, a colossal explosion devastated the Bavarian village of Oppau, part of the vast chemical works centered on Ludwigshafen. Four-and-a-half thousand tons of ammonium nitrate, which had been accumulating for use as fertilizer for many years without mishap, suddenly ignited.[30] The

[28] Seismologists often refer to the P- and S-waves as the 'push-pull' and 'shake' waves respectively.
[29] See Howell 1990.
[30] Jeffreys 1924, pp. 169, 179.

atomic physicist Walter Elsasser, then a young man of seventeen, was in Heidelberg, some twenty-five kilometers away, dressing for school. "Suddenly there was a big, protracted boom as if from a distant explosion, and then my windows opened ever so slowly and steadily as if pushed by an invisible hand. There was a shattering of glass, then silence." A school-teacher arranged a class visit to the site two days later. All was destruction: heaps of rubble, twisted pipes, flattened dwellings. Survivors, some with skin permanently blue or orange from their work at the dye vats, wandered the ruins.[31]

The Oppau was one of the most powerful man-made explosions before atomic bombs. Powerful enough, perhaps, to have been audible in England? Napier Shaw at the Meteorological Office set out to investigate. He found no evidence that the blast had been heard, but turned up smoked-paper seismo-gram traces recorded at several seismic stations that clearly showed resulting shock-waves. Since it was known precisely when and where the 'earthquake' occurred, the traces could be used to make reliable estimates of wave speeds. Wrinch and Jeffreys found the travel times conformed with the hypothesis of waves spreading with uniform velocity in a homogeneous medium, but com-puted speeds of order a few kilometers per second, lower than usual for the earthquakes in Turner's tables. The Oppau explosion, however, had occurred on the surface, whereas earthquakes typically had deep and uncertain foci in inhomogeneous rock. By considering the times for subsequent reverberations to arrive, Wrinch and Jeffreys showed that the traces were consistent with the various reflected and refracted waves that would result if the Earth's crust above the mantle comprised at least two layers, like the model inferred by Suess from geological data around 1880, rather than the single layer inferred by Mohorovičić from earthquake data in 1909. And by comparing elasticities, they could hazard an identification of the types of rock involved. The Earth's crust was mainly granite with something denser underneath.[32]

4.6.3. New work on probability – John Maynard Keynes

Pearson believed that inverse probability could be used in some cases inde-pendent of the choice of prior probability, and continued to rely on balls-in-urn sampling to derive probability distributions.[33] Broad, though persevering with the problem of induction, had tracked it from probability theory to the condi-tion of the universe, and had eventually concluded that induction must work in

---

[31] Elsasser 1978, pp. 20–1.
[32] Wrinch and Jeffreys 1923a.
[33] Pearson 1920a.

practice because human language had evolved into a good approximation of the correct causal categories operating in the universe.[34] Yet there were new developments in probability theory. Most notable was John Maynard Keynes's extended work on logical probability, *A Treatise on Probability*, published in 1921. An extension of an earlier King's College Fellowship dissertation, the *Treatise* was the first major conspectus of probability since Todhunter's history, and with its discursive and accessible style was quickly influential.[35] Like that of Broad and Wrinch, Keynes's definition and mathematical development of probability as a logical concept followed that of W.E. Johnson, his colleague at King's and former tutor. Though Keynes's interest in probability was perhaps not wholly philosophical – Virginia Woolf's diaries record him as a successful gambler, often emerging £40 or more to the good after a weekend session of cards at the Asquith's[36] – he was much more interested in the logical manipulation of probabilities than their quantification. Like many of the nineteenth-century philosophers, Keynes regarded the probability calculus as a broad sort of logic applicable generally to uncertain inference rather than a workable numerical tool. Some probabilities he believed could not be quantified at all.[37] Most of his illustrations in the *Treatise* were thus of ethical or philosophical matters, such as legal testimony. The few scientific examples were either old hat – distribution of star clusters, for example – or trite; psychic research, according to Keynes, was the most important province of probability.[38]

---

[34] Broad decided that induction was, as Russell had claimed, simply a form of deduction in disguise, and that inductive mistakes occur as a result of man's imperfect hold of the correct causal quantities. Numerical probabilities cannot be assigned in these situations. "The certainty of the most certain inductions is thus relative or hypothetical, and the probability of the hypothesis is not of a kind that can be stated numerically." But Broad was by no means convinced that his paper on the subject was the final word: "I am painfully aware that this article is complex and diffuse without being exhaustive. There is hardly a line in it which I could seriously defend even against myself if I chose to be a hostile critic. But I print it in the knowledge that . . . its very badness may convince the charitable reader at least of the extreme difficulty of the subject" (Broad 1920, pp. 44–5).

[35] Broad regarded it as the best work on probability; the young I.J. Good read it "religiously," even while queuing for shows. (Broad 1922; IJG.)

[36] Nicolson 1976, p. 385.

[37] Keynes maintained that degrees of belief are only partially ordered. Some are quantifiable; others had magnitudes but were not fully comparable; still others applied to arguments and reflected the truth of a proposition, but in vague ways.

[38] Economists have debated the extent to which Keynes's thoughts on probability influenced his later work. Carabelli has argued that the concept of probability is central to a full understanding of the economics, and has drawn parallels between Keynes's probability and that of Wittgenstein as expressed in the *Philosophical Investigations*. Cottrell, on the other hand, has downplayed the connection, and instead has attributed such recent moves by

Jeffreys and Keynes, though both at Cambridge, were based in different colleges and had little contact. (They met only once, while later sharing a railway carriage to London.) Yet with the common influence of Johnson, Keynes's approach was broadly similar to that of Wrinch and Jeffreys. Probability, a primitive concept, is a logical property of propositions conditional on data. Though subjective in the sense that it measures rational belief, a probability is objective in having only a single value for each combination of proposition and data, regardless of who is considering it.[39]

Unlike Wrinch and Jeffreys, however, Keynes took an uncompromising stance on the Principle of Insufficient Reason. In part, he was reacting to its promiscuous use by the Victorian economist William Stanley Jevons. Like Galton, Jevons was an enthusiastic believer in the quantification of social variables, and held that personal qualities, not only matters of judgment, but of emotion too, could be measured precisely. Values of pleasure, pain, and desire could be combined into utility functions, and since man was rational, these would lead to economic laws of the same status of natural laws. Keynes, in contrast, regarded man as an imperfect creature of 'deep and blind passions,' and spoke of the "insane and irrational springs of wickedness in most men."[40] It is precisely because people did not act rationally and predictably

economists to embrace Keynes's views on probability to (mistaken) professional concerns. Keynes's research on probability started around 1906, well before his economics. Although the *Treatise* did not appear until 1921, it was largely complete before the war – Broad and Russell went over the proofs in the summer of 1914 – and followed a 1911 Royal Statistical Society paper on estimation. The notion of uncertainty that appears in his economics is broadly subjectivist. Investors must assess the probability of return conditional on news of the market, for example, and the preference for cash holding depends on expected interest rate changes. Frequencies are not all that easy to come by for the individual investor in economics. Yet only strict and restrictive frequency definitions, such as Keynes's own or that of von Mises, would rule that economic events, as products of imprecisely specifiable and changing conditions, were inapplicable for a frequency treatment. (A proper *kollektiv* in von Mises's sense might not be available, but an infinite hypothetical population certainly is.) More relevant is Keynes's view that probabilities are often incalculable. Significantly, Keynes rarely referred to the *Treatise* in later economic writings (Carabelli 1988; Cottrell 1993; Broad 1922, p. 72; von Mises 1957; see also Conniffe 1992 for the links between Keynes's and Fisher's versions of probability).

[39] Keynes thought that the ability to perceive the values of logical probabilities was intuitive, though he conceded that the intuitions of some men were more reliable than those of others. He was also occasionally reluctant in his 'objectivism.'

[40] Keynes implied these comments, in his 1938 essay, "My Early Beliefs," to his views following WWI, by which time he had repudiated the naive pre-war belief of the Bloomsbury group in "a continuing moral progress by virtue of which the human race already consists of reliable, rational, decent people, influenced by truth and objective standards," and instead recognized that civilization was sustained on "a thin and precarious crust erected by the personality and the will of a very few." See Keynes 1949, pp. 83–7, 98–9, 102.

that it was necessary to take out insurance, and insist on contracts in business deals. Jevons was a menace to economics. And his wild applications of the Principle of Insufficient Reason to human behavior Keynes held responsible for the poor reputation of inverse probability. In the *Treatise*, therefore, Keynes presented a restricted but safe version of inverse probability that did not depend on the Principle of Insufficient Reason (or, as he called it, the Principle of Indifference). To strengthen the case for this new formulation, he devoted considerable space to ridiculing the classical exposition of Laplace, from whom, via De Morgan, Jevons had derived his probabilistic approach. Keynes's citing Bobeck's calculation on the rising of the sun as a counter-example to the Rule of Succession is a case in point. He also drew from von Kries and the 'fatal' criticisms of Boole and Venn.[41]

Jeffreys thought that Keynes's attitude toward the Principle of Insufficient Reason stemmed from a confusion over the relevant background knowledge.[42] Keynes had rejected Jevons's view that with no information, the probabilities of each of two exhaustive and mutually exclusive probabilities is $\frac{1}{2}$. Ignorant about the color of a book, we should according to Jevons conclude that the probability of both statements "this book is blue" and "this book is not blue" is $\frac{1}{2}$. Keynes argued that since the same could be said about other colors – red and yellow, say – Jevons's position would force the contradiction that the probabilities of each of "this book is blue," "this book is red," and "this book is yellow," is $\frac{1}{2}$. Jeffreys countered that the statements in each case are based on different evidence, here concerning recognizable colors. Since the probability of a proposition is always relative to a given body of data, it is natural that its value will change when we change the relevant data. If we can recognize the colors blue, red, and yellow, we should properly regard the probability of each as $\frac{1}{3}$. If we can recognize only blue, then the probability of both "this book is blue" and "this book is not blue" is $\frac{1}{2}$. It was this misconception, thought Jeffreys, that lay behind Keynes's caution in matters like sampling, and his reluctance to quantify probability, in direct contrast with Wrinch and

---

[41] Keynes extended his derision from Laplace's formulation of probability to his entire mathematical edifice, additionally representing him as an obsequious lickspittle in accepting the post of Minister of the Interior from Napoleon. (Conniffe ventures that Keynes would have been still more scathing had he known what a poor fist Napoleon thought Laplace had made of this job.) Indeed, Keynes suggested that the whole of the classical French school was obsolete and should be replaced by the mathematically heavyweight Germans and Russians. In reply, Broad noted: "I am afraid that, with the exception of Lexis, these names are mere sternutations to most English readers; but I suppose we may look forward to a time when no logician will sleep soundly without a Bortkiewicz by his bedside." Conniffe 1992, p. 480; Broad 1922, p. 81.

[42] Jeffreys 1922.

Jeffreys's axiom that all probabilities could be put in a precise numerical order.[43] Keynes had wound up concluding that Bayes's theorem, although the proper way to manipulate numerical probabilities, could only be mobilized when the possible alternatives were expressible in the same verbal 'form' and supported equally by the background information. Since relatively few probabilities could be quantified in any case, and since increasing quantities of data are *less* likely to balance exactly across the alternatives, probabilistic inference was therefore generally not a quantitative matter.[44] Jeffreys regarded such a version of numerical probability theory, though doubtless interesting from a logical point of view, as barren because inapplicable in most practical situations. Nevertheless, despite these reservations, Jeffreys approved of the *Treatise*. He regarded Keynes's critical stance toward the frequency definition as particularly praiseworthy. Keynes had argued that such a definition applied only in definite and unchanging circumstances, and required complete knowledge before probabilities could be assessed; yet, as he was later to say, our only knowledge of the limiting case is that, "in the long run, we shall all be dead."[45]

---

[43] Frank Ramsey later converted Keynes to the view that all probabilities could be quantified. See Keynes [1963], pp. 243–4.

[44] Keynes 1921, p. 176. See Broad 1922.

[45] See Keynes 1921, pp. 92–110 for his criticisms of the "widely held" frequency theory, which he dismissed as too quantitative and precise to apply generally to epistemic uncertainty. Venn's frequency-based probability statements had little use if applying only to collectives, and were arbitrary and unconvincing if transferable to individuals: to argue "from the mere fact that a given event has a frequency of 10 per cent in the relevant instances under observation, or even in a million instances, that its probability is 1/10 for the next instance . . . is hardly an argument at all." Keynes 1921, pp. 407–8 (see too Conniffe 1992). Unsurprisingly, Fisher was caustic in his 1922 review of the *Treatise*. He criticized Keynes's "psychological" definition of probability, his distasteful sneering at the efforts of previous probabilists and statisticians, his unfamiliarity with modern statistical theory, and the mistakes littered through the text. ("It is difficult to discover any numerical example which appears to be correct.") He also demurred at the extravagance of Keynes's antagonism toward the Rule of Succession. To Keynes's complaint that the precise probability statements conjured from the rule stood in contradiction to the starting assumption of ignorance, Fisher replied that the form of 'ignorance' assumed in deriving the rule – the Principle of Insufficient Reason – implied that the starting probability of drawing a black ball from an urn filled with only with black and white balls was $\frac{1}{2}$, exactly that given initially by the Rule of Succession. "In the present writer's opinion the assumption of equal distribution is usually illegitimate, but it involves no such inconsistency as Mr Keynes imagines." Thirty-three years later Fisher was still defending the Rule of Succession, describing Venn's attacks as "rhetorical sallies intended to overwhelm an opponent with ridicule." Zabell attributes this unexpected and ongoing support to the fact that the Rule follows from the fiducial argument. Fisher 1922b, p. 50; Fisher 1956, p. 25; Zabell 1989b, pp. 250–251.

4.6.4. Other factors

Wrinch and Jeffreys's conviction that probability was the key to the process of induction was further strengthened in early 1921. Before leaving for China in 1920 for a year as a visiting lecturer, Russell had entrusted Wrinch with the manuscript of Wittgenstein's *Tractatus* and the job of finding it a publisher. (After rejection in England, the manuscript was eventually accepted by the German journal *Annalen der Naturphilosophie* in 1921; Wittgenstein regarded the error-strewn article as a 'pirate edition.'[46]) Wrinch, who revered Wittgenstein – indeed, experienced anxiety dreams at the prospect of meeting him – found much in the *Tractatus* to support and reinforce her view of science and interpretation of probability.[47] The Great Man, who clearly had little truck with the frequency interpretation, adopted a straightforward subjective view in which probabilities were functions of propositions relevant when knowledge was incomplete. He even used the balls-in-urn model to illustrate his point. "Now, if I say, 'The probability of my drawing a white ball is equal to the probability of my drawing a black one,' this means that all the circumstances that I know of (including the laws of nature assumed as hypotheses) give no more probability to the occurrence of the one event than to that of the other. That is to say, they give each the probability $\frac{1}{2}$ as can easily be gathered from the above definitions. What I confirm by the experiment is that the occurrence of the two events is independent of the circumstances of which I have no more detailed knowledge."[48]

Jeffreys, who had undergone psychoanalysis following personal difficulties, meanwhile found in Pearson's description of the unity of science a

---

[46] Monk 1990, pp. 203–5.
[47] For example, on scientific laws: 6.37–6.371, "There is no compulsion making one thing happen because another has happened. The only necessity that exists is logical necessity. The whole modern conception of the world is founded on the illusion that the so-called laws of nature are the explanations of natural phenomena." On the ordering of probabilities: 5.1, "Truth-functions can be arranged in series. That is the foundation of the theory of probability." And on the assignment and the propositional nature of probabilities, see 5.152, 5.153, 5.154, 5.155. (Note, though, that the statement in the first part of 5.154 – "Suppose that an urn contains black and white balls in equal numbers (and none of any other kind). I draw one ball after another, putting them back into the urn. By this experiment I can establish that the number of black balls drawn and the number of white balls drawn approximate to one another as the draw continues." – is strictly false.) Wittgenstein's remarks on probability were influenced by Johannes von Kries, who argued in 1886 that a probability was subjective but could be estimated statistically. Wittgenstein regarded the 'problem' of induction as only apparent, arising from asking the wrong sorts of question.
[48] Wittgenstein 1922, § 5.154.

justification for the still disreputable science of psychology.[49] Psychology was not a quantitative study, but neither were the majority of the biological sciences. And that psychoanalytic concepts, such as neuroses, may disappear with recognition was perhaps no different than wave functions, or quanta of light, that were destroyed during observation. Jeffreys admitted that the facts of psychology were few. Yet it was a science nonetheless. In gradually incorporating new empirical data into a developing schema of organizing concepts, it was no different from physics. Indeed, here was a connection to probability theory: in stressing the primacy of sensations, Pearson's account was profoundly individualistic.[50] Science was essentially a private matter, and probability, in reflecting an individual's attitude toward propositions, part of psychology. Had not Wittgenstein, too, argued that induction can

[49] Cameron and Forrester suggest that Jeffreys began analysis sometime between 1917 and 1922, while he was living part-time in London and working at the Met Office. James Strachey, a member of the Bloomsbury Group and translator of Freud, briefly met Jeffreys in London during this period, probably as he was accompanying Wrinch on a visit to see Russell. Strachey assumed Jeffreys and Wrinch were engaged, and later speculated that it had been her jilting him that led him to seek analysis (Strachey 1986, p. 223). Jeffreys's first analyst was Ernest Jones, the "self-appointed patron and manager of English analysis." Jones was a disciple of Freud, and later wrote to the Master that Jeffreys combined "the highest intellectual capacities with a very low degree of ordinary social capacities, such as *savoir faire*, common sense, etc. I think the success of the analysis was only moderate. He has an astounding facility for rapidly acquiring the profoundest knowledge of any subject, e.g., botany, higher mathematics, physics, etc., and has made himself world famous through his mathematical researches into physics and the structure and movements of the earth, on which he has written some heavy books." Jones to Freud, 6 July 1936, quoted in Cameron and Forrester 2000, p. 221. See also Jeffreys 1931, p. 204.

[50] "Each sensation of each judgment is absolutely private to the individual.... Any satisfactory account of the scientific method must therefore rest at bottom on the private sensation and the individual judgment," wrote Jeffreys in a 1921 piece criticizing the theory of science propounded by the physicist Norman Campbell in his book *Physics, the Elements*. Like Jeffreys, Campbell had taken an early role in the National Union of Scientific Workers, and his view that scientific practice required no mathematical or logical justification shared many features with Jeffreys's positivism. The two differed sharply over the role of probability in science, however. Campbell considered scientific inquiry to be restricted to those areas in which universal agreement about experimental results obtained, and consequently regarded probabilities as wholly inapplicable to propositions or hypotheses. Furious at what he saw as misrepresentation of his views, Campbell engaged in a brief but agitated correspondence with Jeffreys in late 1921. ("You say that my definition of probability is scientifically inapplicable; if by that you mean inapplicable to the probability of propositions, of course it is; that is why I chose that definition.") No compromise was reached, and as with Jeffreys's dispute with Fisher, this seems in part a result of differing research experiences. "In what you call science," complained Campbell, "I cannot recognize a single feature of the study which is my special interest." (Jeffreys 1921d, p. 569; Campbell to Jeffreys, 21 October 1921; Campbell to Jeffreys, 19 September 1921; HJ.)

only be justified psychologically?[51] Jeffreys was intrigued by *Instinct and the Unconscious*, a popular introduction to psychoanalysis by his St. John's colleague, the psychologist and anthropologist W.H.R. Rivers. The description of the individual's learning process was particularly suggestive. Jeffreys wrote Rivers that his explanation of the formation of beliefs was "a good example of the method of successive approximation that Miss Wrinch and I have been saying is a necessary characteristic of science."[52]

4.7. PROBABILITY THEORY EXTENDED

4.7.1. The Simplicity Postulate

Wrinch and Jeffreys's view of the significance of inference in science had evolved during the debate on relativity. Though their call for a restrained approach to Einstein's theory was largely ignored – few physicists preferred their prosaic, practice-first approach to the opportunity for elevated philosophizing[53] – they grew committed to the view that science was based on action, and characterized by the gradual building of laws by inference.[54] A theory of science must naturally include the purely logical manipulations that governed the deduction of testable observations from a law or hypothesis, but more essential was the primary process of generalization from the data of the senses to descriptions of the world. Such generalizations were always tentative: as Einstein had shown, even laws as secure as Newton's could be overturned. In the paper of 1919, Wrinch and Jeffreys had considered probabilistic induction in the restricted case of sampling. By 1921, however, they had come to insist that the calculus of probability was not merely applicable to probabilistic laws, but to the deterministic laws that made up the bulk of scientific knowledge. Since such general laws were at best approximations, constructed by inference from sense data, they were to be assessed with the calculus of probabilities. A theory of probability, therefore, applied to the whole of scientific practice.

---

[51] See Wittgenstein 1922, § 6.363 and 6.3631.

[52] HJ (Jeffreys to Rivers, 8 January 1922). Jeffreys corresponded occasionally with Rivers, and the two became quite friendly by Rivers's death in June 1922.

[53] See Jeffreys 1933a, p. 526.

[54] This version of positivism was similar to the operationalist philosophy that the American physicist Percy Bridgman was developing around the same time as a response to the relativity controversy. Jeffreys subsequently corresponded with Bridgman, and favorably reviewed his 1927 book, *The Logic of Modern Physics* (Jeffreys 1928).

Jeffreys and Wrinch soon started to refer to general scientific laws with the language of probabilities. Their paper on mensuration argued that the concept was essential in order for measurements to be extracted from a combination of sense data, and that Euclid's postulates should be viewed as at best an inference of high probability. Considering alternative explanations for Eddington's data and the perihelion of Mercury, Jeffreys pointed out that it is not enough that a theory account for the observations; to be probable it must be the only theory so to do. Likewise, he advised that superfluous concepts like 'space' were to be avoided because the "introduction of additional hypotheses decreases the probability of the theory."[55]

Quantifying such probabilities, however, was a different matter. Though many philosophers and scientists of the late nineteenth century had regarded inverse probability as a model of scientific inquiry, few had attempted to put the theory into a mathematical form that could be profitably used by scientists. Evaluating the prior probabilities of hypotheses was the major problem. Wrinch and Jeffreys had disentangled the assignment of priors from the formal development of the mathematical theory of probability in their paper of 1919, but had conceded that prior probabilities were nevertheless rarely easy to determine. This previous discussion, however, had considered general laws of the form 'all crows are black.' From their experiences in geophysics, they had since come to see that the classical view was misleading. Few laws encountered in science were enumerative generalizations concerning events, to be modeled by a process analogous to drawing colored balls from an urn. Instead, they were typically functional relationships between numerical quantities, and were inferred usually by fitting polynomials to combinations of measurements. Thus seismologists typically scrutinized various types of incoming data from seismology stations and tried to reduce them to quantitative relations. And *pace* the 1919 paper, scientists wished to assess directly the probability of such general laws – rather than just the probability of the next draw in the enumerative case – since they often wished to extrapolate beyond the original data.

Finding prior probabilities for general laws of this sort, however, introduced the additional problem, also raised by Broad in his 1918 paper, that a finite set of data points could be fitted by an indefinite number of quantitative laws. With an infinity of allowable laws, the prior probabilities of each must be set effectively to zero. Since Bayes's theorem states that the posterior probability of a hypothesis is proportional to its prior probability, there is then nothing for the inverse mechanism to get a purchase on. Not

---

[55] Jeffreys 1921a, p. 394.

only could a general law never attain a high probability, it can never be anything but extremely unlikely. Like Pearson, Broad had seen that non-uniform prior probabilities would solve the problem within a framework of inverse probability. Yet with his idealistic faith in the hypothetico-deductive model of science and the Principle of Insufficient Reason, he was reluctant to grant that some hypotheses might be privileged over others.[56] Jeffreys, whose theoretical studies were based on the experimental spade-work of others, knew however that the practice of science required the acceptance of testimony. He realized that laws of the form 'all crows are black' are believed not solely on the basis of limited observations, but because these observations do not contradict a prior judgment already believed to be highly probable. Even so, in the general case, each admissible functional relation must start with a finite and non-zero prior probability, because scientific laws are always defeasible. How could this be justified if there were an indefinite number of candidates?

The first piece of the puzzle came to Wrinch during a picnic lunch taken together on Madingley Hill. Fixing numbers to propositions had been central to the 1919 paper. Could the idea perhaps be extended to general laws? Her work with trans-infinite numbers suggested an approach. Assume that "[e]very law of physics is expressible as a differential equation of finite order and degree, with rational coefficients."[57] Then all establishable or demonstrable quantitative laws must form an enumerable set. That is, they could be ordered against the positive integers. So each law could correspond to a term in an infinite but convergent series. With the sum of the series normalized to unity, each term in the sequence can be identified with the prior probability. Thus even with an infinity of possible physical laws, potential candidates need not start with infinitesimal probabilities.

The assumption, they reasoned, seemed plausible: most constants disappear when a law is reduced to the form of a differential equation; continuous variables such as mass, electric charge, size, and so on might in any case be discrete, according to modern ideas on the discontinuous structure of matter. But it remained to be seen whether such laws could be ordered in a consistent way. Jeffreys turned to the *Grammar*. Pearson's view of scientific practice was somewhat severe – frequently one wished to infer beyond the data, and

---

[56] Responding to the problem of general laws, Keynes had followed Broad with an appeal to some extra-probabilistic feature of the universe, a premise that somehow expressed the 'uniformity of nature.'

[57] Wrinch and Jeffreys 1921a, p. 386. See also HJ-DVL; HJ (undated note [1978] for a letter to Zellner). Jeffreys later justified the assumption that allowable laws form an enumerable set by pointing out that all hypotheses that scientists might want to test can be stated in a finite number of words.

an entity surely did not need to be directly observed to be useful – but his brand of positivism was on target when it came to the scientific method as more description than explanation. Wrinch and Jeffreys were attempting to construct a science of science. Hence they should be aiming not at the justification of science, but simply its formal and concise description, constructed from observations of good practice and rational judgment.

How then did real scientists treat general laws? Wrinch and Jeffreys recalled their choice between the various mathematical forms on offer in fitting their seismological data. They had applied Occam's razor and chosen the lowest-order polynomial possible. This was not merely a matter of convenience, of preferring a simple law over a more accurate but complicated form, since they had confidence in their inference that the Earth's crust was made up of two layers. Nor was their curve-fitting an exercise in short-hand description, since the inferred travel-time relationships could be extrapolated beyond the limits of the original data to make predictions and test hypotheses about the Earth's core. The simple relationships instead seemed to have some weightier status. Indeed, Wrinch and Jeffreys had selected a simple equation even though a higher-order polynomial could have fit the data better. Unlike the counts or events of the enumerative model, genuine scientific laws were inferred from measurements that were never exact fits. Yet Wrinch and Jeffreys had been prepared to sacrifice accuracy of description to adopt a simple relationship. Further, this commitment to the simple law was such that the differences between the data and the corresponding values from a fitted linear or quadratic curve were ignored as "errors," even though observations were fundamental in the positivist model of science, and strictly could never be regarded as 'wrong.' And this preference for simplicity was not confined to quantitative laws. Jeffreys had remarked during the relativity debate on the common assumption that a physical law must be simple in form. Eddington's discussion of space and the aether, for example, was based on the tacit assumption that it must be mathematically tractable.[58] Broad too had gestured at the importance of simplicity for qualitative laws.[59] That scientists prefer simple laws is an empirical fact; it can provide a basis for ordering prior probabilities. Wrinch and Jeffreys announced a 'Simplicity Postulate': the

[58] Jeffreys 1921a, pp. 267–8.
[59] With his commitment to the Principle of Insufficient Reason, Broad had embarked on the flimsy argument that complex hypotheses tend to have lower probabilities because they are really unique combinations of simpler, and equally likely, independent unit propositions. Recognizing that the argument was probably spurious – there is no unique way to break down a hypothesis into a set of propositions; even if there were, the 'tendency' is only true if propositions, by and large, have similar and independent probabilities – he did not pursue it. Broad 1918, p. 402.

simpler the law, the greater its prior probability. The Bayesian machinery can finally be cranked up.

4.7.2. The papers of 1921 and 1923

Wrinch and Jeffreys developed these ideas in a paper, "On certain fundamental principles of scientific inquiry," published in the *Philosophical Magazine* in two parts in 1921 and 1923. They began with another attack on the frequency theory of probability. Science is characterized by the inductive generalization of laws or theories from the facts of sensory experience. This is clearly a probabilistic process, yet since the properties of infinite series necessary for the frequency interpretation are themselves inferences, probability must be regarded as a primitive concept that reflected a degree of knowledge rather than any sort of frequency.

The new theory of probability was not simply a subset of logic or a way to treat chance events, but to be regarded as a model of scientific inquiry.[60] And since in Pearson's scheme the scientific method was simply a more formal and public expression of the commonsense learning experiences of the individual, Wrinch and Jeffreys introduced their work as a theory of knowledge. ("The theory we are attempting to construct is one that includes the processes actually employed by scientific workers; since psychology is by definition the study of behavior, this work may perhaps be regarded as a part of psychology."[61]) Such a theory was itself part of science, they argued, and should be studied by scientists. It was because the study of science had been hijacked by philosophers, who, obsessed with intensive analysis and the rules of logic, had completely missed its essentially inferential character, that the scientific enterprise was still obscure.

Wrinch and Jeffreys went on to embrace a full-blooded interpretation of inverse probability, and showed how Bayes's theorem could account for many aspects of intuition and good scientific practice.[62] The presence of prior probabilities was no embarrassment, since priors represent the expertise or previous knowledge a scientist brings to the data. ("Now it appears certain that no probability is ever determined from experience alone. It is always influenced to some extent by the knowledge we had before the experience."[63]) And that the law "established with a high probability by experience is never

---

[60] The title of the 1921 paper, "On certain fundamental principles of scientific inquiry," compares with that of the 1919 paper, "On some aspects of the theory of probability."

[61] Wrinch and Jeffreys 1921a, p. 372.

[62] See Appendix 2.

[63] Wrinch and Jeffreys 1921a, p. 381.

an approximation to the simple law, but the exact simple law itself," licensed "extrapolation over an indefinitely wide range . . . with the full probability of the law."[64]

Yet as Wrinch and Jeffreys made clear, they were not attempting to justify or explain scientific reasoning. Quite the reverse: the Simplicity Postulate was introduced not as a logical response to the problem of priors, but as a plausible proposition necessary to bring the probability calculus into line with the habitual and successful practice of scientists working with simple quantitative relations. The goal of their theory was simply to help "the understanding of the structure of physics, and thereby to suggest which methods of development and criticism are in closest accordance with the principles that are generally believed to lead to reliable results."[65] Philosophical methods were not totally worthless. Jeffreys and Wrinch read Whitehead and Russell as working backwards from mathematical practice to a set of consistent axioms, and decided that their own theory should likewise be pared and tuned until it could most economically coordinate scientific practice. But they wished to expose the fallacy beloved of philosophers that a theory of knowledge could be 'proved' independently of empirical experience. Their axioms were not proved, merely plausible to the primitive starting-point of commonsense. "There is no question of making unjustified assumptions; for we make any assumption that we believe correct, and then work out its consequences . . . we are assisting common-sense to build up a consistent and comprehensive attitude, and that is the aim of science."[66]

A consistent attitude could also to serve as a guide to practice. Though simplicity was a quality more easily recognized than quantified, some progress was possible.[67] To start, differential equations of low order and degree were, *ceteris paribus*, presumably more simple than higher-order equations. Consider laws of the same form, differing only in the values of their constants. With each constant expressed as a reduced fraction, it seems natural that simplicity be related to the sum of numerator and denominator. Hence laws featuring constants that differ slightly from an integer will have far lower prior probabilities than those with integers. So Asaph Hall's suggestion that the

---

[64] Wrinch and Jeffreys 1923b, p. 370.

[65] The Simplicity Postulate "therefore constitutes a contribution to scientific methodology, which aims at reducing the propositions habitually assumed in scientific investigations to groups of more fundamental propositions which imply them, and is not an attempt to prove that these fundamental propositions have themselves a basis in logic" (Wrinch and Jeffreys 1921a, p. 372).

[66] Wrinch and Jeffreys 1921a, p. 378.

[67] Wrinch and Jeffreys 1921a, p. 386.

perihelion of Mercury could be explained under Newton with a 2.000000016 gravitational power law of the Sun could be regarded as so extremely improbable that it would require enormous quantities of verifying data to lend it any weight. It was this, stated Wrinch and Jeffreys, quite apart from the negative empirical results actually obtained, that led to the proposal's rejection.[68]

4.8. THE COLLABORATION STARTS TO CRUMBLE

Both Wrinch and Jeffreys had found their way back to Cambridge by the publication in 1923 of a further co-authored paper, developing their earlier discussion of mensuration as an extensive procedure built by probabilistic inference from sense data.[69] Jeffreys had become disillusioned at the Meteorological office after the Armistice. He had been shifted to take charge of the library in 1920 by Shaw, who was preparing to retire, and who tried to sweeten the position with the suggestion that it include the larger role of reorganizing the collection rather than just the duty of cataloging new additions. But with less freedom to pursue his own interests, Jeffreys found himself occupied answering tiresome requests for weather records. (These, it subsequently turned out, were crucial for court cases concerning road accidents.) It was time to go. Jeffreys was anxious to find a permanent academic post, but following a cursory scouting expedition in London, decided he would be better served returning to Cambridge. Though he had taken nearby lodgings while at the Met office, Jeffreys had commuted back to college at the weekends. Having persuaded the Senior Tutor that his small undergraduate set of rooms in Chapel Court was no longer sufficient, he resigned his post at the Met Office early in 1922 and returned to John's as college lecturer in mathematics.[70]

Wrinch, though, had been the first to leave London. Her circumstances had changed abruptly in 1919 after Russell became romantically involved with Dora Black, a firebrand feminist and socialist. Wrinch had been at Girton with Black, and indeed had introduced her to Russell in 1917. But as he became besotted with the young radical, so Wrinch's close relationship with him waned. She moved back to Girton in 1920, and when Russell and Black married in 1921 found herself abandoned. Mixing with the social and intellectual aristocracy had been intoxicating for the daughter of a waterworks

[68] Wrinch and Jeffreys 1921a, p. 389.
[69] Wrinch and Jeffreys 1923c. Though co-authored, the mensuration paper was largely Jeffreys's work. (As in the earlier papers, the order of authorship reflected chivalry rather than relative contribution.)
[70] BJ, HJ-MM, HJ-DVL.

engineer. Yet friendship with Russell would not guarantee a job. As a woman, establishing herself in academe was going to be difficult. And the period in the vanguard had left her with modish interests that straddled the conventional disciplines. In 1922, and with an eye on her career prospects, Wrinch married John Nicholson, a recently-appointed fellow in mathematics and physics at Balliol College, Oxford. Nicholson was thirteen years her senior; she had attended his lectures in her first year at Girton. She continued to work with Jeffreys while both were in Cambridge, but after Nicholson found students for her in Oxford, she joined him there as an academic, leaving Girton in 1923 and effectively ending the collaboration.[71]

By this time, Jeffreys too was experiencing career difficulties. His college fellowship did not give security of tenure, and the offer from Harlow Shapley of a one-year post at the Harvard Observatory, together with vaguer noises from the British-born astronomer and mathematician Ernest Brown at Yale, was sufficiently tempting for him to sail to America in 1923 to investigate. The collaboration with Wrinch was uncharacteristic: Jeffreys was reserved by nature, and awkward in company, and had chosen research fields and methods that allowed him to work almost entirely alone – typically with his typewriter on his knees, his hand-cranked Marchant calculating machine on the floor in front, and the room ankle-deep in research papers and works-in-progress. With no administrative or academic duties, the Harvard post seemed tailor-made. He could devote himself to theoretical research, flitting freely between the topics that interested him.[72] Yet Jeffreys valued his college tutorials. Though notoriously unprofitable for the students – the geophysicist E.R. Lapwood recalled the embarrassment that followed Jeffreys's silent and rapid completion of unsolved weekly assignments – Jeffreys found that contact with undergraduates provided an invaluable mental stimulus to tackle problems in new ways. Gambling that the ongoing Royal Commission on Cambridge would make permanent academic posts easier to come by, Jeffreys declined Shapley's offer.[73]

[71] Abir-Am 1987, pp. 247–50; B. Jeffreys 1991, p. 22. Lady Jeffreys, then an undergraduate at Girton, recalls seeing Jeffreys cycling up to see Wrinch around 1921.

[72] Jeffreys's output was prodigious, even when compared with that of contemporary scientists, for whom academic promotion depends more on quantity than quality of work. He published his first paper in 1910 aged 19, and his last almost eighty years later in 1989, the year of his death. Alan's Cook's bibliography, which is not complete, lists more than 440 papers published in between, only a handful of which are co-authored. Jeffreys's collected papers fill six volumes.

[73] B. Jeffreys 1991, p. 22; B. Jeffreys 1992, p. 305. See too BJ; Lapwood 1982.

Jeffreys was industrious during this period of uncertainty. The common thread of his work remained the evolution of the solar system. The evidence seemed to support his tidal theory; at least, this theory seemed to account for more facts than it supposed. Yet inferences from the astronomical data alone were infuriatingly vague. In astronomical terms, the Earth is no more interesting than any other planet. But since data on the constitution of the Earth was relatively varied and plentiful, Jeffreys decided that a focus in geophysics would be the best way to study the origins of the planets. His broad conspectus of the field, *The Earth*, published in 1924, was immediately successful.[74] The combination of diverse effects from geology, astrophysics, and geophysics brought to bear on the physics of the Earth as a whole exemplified Jeffreys's style. The compression forces required to create mountain ranges, for example, could be estimated from geological values of internal strength and combined with a model of thermodynamic cooling to yield estimates for the initial temperature and composition of the Earth. These, in turn, could be related to values of its age calculated from tidal retardation or the radioactive effect, and cross-checked with the deceleration estimated from the dynamics of the Earth–Moon system or the residual equatorial bulge, or needed to account for the formation of the oceans and atmosphere. Finally, these values could be equated to eclipse records, the observed eccentricity of Mercury, or the frictional slowing of the planets through the gas of condensation, and so tied in to a grand model of the solar system.

Publication of *The Earth* marked the beginning of Jeffreys's commitment to geophysics. His specialty became seismology. As the first to use an artificial explosion as a seismic source, the work with Wrinch on the Oppau explosion gave notice of the potential power of seismology to probe the interior of the Earth. Yet the Oppau work also clearly demonstrated the need for improved time-travel tables if such relatively shallow effects could be extrapolated deep down into the crust. Jeffreys set about combining and analyzing the time-travel data from seismological stations around the world. These observations could be used to refine his models and adjust their parameters: the Earth is not really homogenous and elastic, but includes cusps, discontinuities, and velocity gradients. The discipline of seismology was still

---

[74] Running to many editions, *The Earth* inspired a generation of geophysicists. E.R. Lapwood, for example, bought the book as a college prize and found for the first time "not the artificial problems of schoolboys' 'applied mathematics,' but subjects arising from the real world of modern science." Lapwood 1982, p. 69.

dominated by field workers and amateurs, and Jeffreys's application of rigorous theoretical and statistical techniques soon made a mark. Not all the older men welcomed his contribution. For example, Jeffreys differed with Turner's 1922 value of 200 km or so for a typical earthquake depth, estimated from travel-time discrepancies in Japanese quakes. Instead, he argued from a theoretical discussion of the characteristic vibrations on a taut string, that the oscillations of shallow seismic waves cannot be generated by deep foci. Though withdrawn and uncommunicative, he was confident enough in his models to publish contentious or heretical results, and proved forceful and insistent when challenged.[75]

Jeffreys's most notable success concerned the Earth's core. Wiechert had proposed a dense center of the Earth in 1905 to account for the slight flattening of the Earth at its poles. The seismic evidence of Oldham around 1906 and later Gutenberg – in particular that the velocity of P waves traveling through the Earth decreases, while S waves are absent altogether – indicated that a clear boundary existed in the center. In 1926, Jeffreys calculated that the Oldham–Gutenberg boundary was a discontinuous increase in density, and that it corresponded to Wiechert's distinct dense central region. Characteristically, he brought additional data to bear. The gravitational forces of the Sun and Moon produce an elastic vibration in the Earth, the period of which gives a value for the global, or average, rigidity of the Earth much lower than that given by shear wave velocity considerations for the mantle. Taking into account the equatorial bulge of the Earth due to its spin, and including compression and contraction effects, Jeffreys was able to show that these external torques are consistent with the rates of nutation of the gyroscopic motions only if the core region is of negligible rigidity. Until 1925, most geophysicists discounted the evidence of the absence of S waves – which, as shear motions, can only travel in solids – and assumed with Lord Kelvin that the Earth was entirely solid. Jeffreys's inference, however, was not only that the Earth's core was a distinct region, but that this core was a dense liquid, probably molten iron, perhaps with some nickel too.[76] With such successes went

[75] See Jeffreys 1923a. Jeffreys regarded this as a purely mathematical matter, and Turner's harmonic analysis as nonsense. He attempted to interest Pearson in the debate, but the older man, in his only contact with Jeffreys, declined involvement. Jeffreys's analysis was vindicated in the late twenties by direct observation of local earthquakes. (Modern understanding is that the focus of most earthquakes is less than 15 km deep, see Bolt 1991, p. 17.) Jeffreys also had the better of the American astronomers Forest Moulton and Thomas Chamberlain, with whom he clashed over their support of a planetismal theory of the solar system and attack on the tidal theory. See Jeffreys 1929 and preceding editorial comments.

[76] Jeffreys 1926. See Brush 1996, pp. 144–7, 187–94.

professional advancement. Jeffreys was elected Fellow of the Royal Society in 1925, university lecturer in 1926, and Reader in geophysics in 1931.[77]

4.10. PROBABILITY AND LEARNING FROM EXPERIENCE – *SCIENTIFIC INFERENCE*

4.10.1. Science and probability

By the end of the twenties, Jeffreys was becoming known both as an inventive and productive theoretical geophysicist, and as a scientific author. *The Earth* had been a great success (and was updated in 1929 with an expanded account of seismology). He followed it in 1927 with a book on operational methods – extending the calculus of the misunderstood Oliver Heaviside – and started to plan a book on tensors.[78] His concentration in geophysics and astrophysics had kept his attention on problems of scientific method. These sciences rely on observations rather than experiments, and the relevant data was often scarce and of variable quality. Yet laws are surmised and extrapolated over enormous ranges. His tidal model of the solar system hung together on Newton's laws, friction, and thermodynamics. How confident could one be of such judgments? In the papers with Wrinch, Jeffreys had considered such issues of principle in general terms, and in an appendix to *The Earth* he had used inverse probability to argue qualitatively that his tidal theory of the solar system was likely.[79] Yet his increasing focus on seismology made

---

[77] Jeffreys's other work around this time was heavily calculational. Examples include his ongoing studies of nutation in the Earth–Moon system, and his pioneering work on the circulation of the atmosphere (from which he showed toward the end of the 1920s that cyclones are an important part of atmospheric circulation, and balance the angular momentum carried to or from the Earth from winds). Some of these problems required wholly new techniques. Investigating the behavior of Mathieu functions while modeling oscillations on an elliptical lake in 1924, Jeffreys developed a method for the approximate asymptotic solution of certain second-order ordinary differential equations. The form corresponds to a series expansion of approximate solutions to the Schrödinger equation, and became known as the WKB technique after independent derivation by Wentzel, Kramers, and Brillouin. See Cook 1990, pp. 310, 316–8, 326–7; Lapwood 1982; Knopoff 1991.

[78] Jeffreys 1927. The amateur Heaviside was the victim of prejudice from professional mathematicians. See Hunt 1991a. Jeffreys had been introduced to his operational methods by Bromwich while an undergraduate at St. John's, and tightened his commutation and convergence conditions.

[79] With a number of free parameters to stand for various items of background evidence – knowledge of the laws of physics, the present condition of the solar system, etc. – Jeffreys manipulated the equations of inverse probability to conclude that "given the empirical data of physics and the existence of the solar system as it is, it is practically certain both that the laws of physics used in the argument are true and that the initial conditions required by

him aware of the practical need for definitive numerical criteria for the evaluation of hypotheses and the probabilistic combination and reduction of data. Though he had lost contact with Wrinch, Jeffreys decided their work should be expanded into a unified account of scientific method. *Scientific Inference*, the result, was published in 1931.

4.10.2. Scientific Inference

Jeffreys began his book with a streamlined version of the work with Wrinch. Science is not a logical process of deduction from *a priori* postulates, because general statements are never known to be true. Instead it is based on the inference of general laws from a set of observed facts. Jeffreys presented the turmoil of the rejection of Newtonianism and introduction of quantum theory as a clear demonstration that even well-established laws are only provisional, to be asserted not with certainty but with a degree of probability that could be revised with new information. Thus does science progress as experimental anomalies force the refinement of inferred laws, "and probability, from being a despised and generally avoided subject, becomes the most fundamental and general guiding principle of the whole of science."[80]

Jeffreys went on to derive the mathematics of probability. The Theorem of Inverse Probability – accredited to Bayes in a footnote, his only mention – he declared is "to the theory of probability what Pythagoras's theorem is to geometry."[81] The 1919 embarrassment over prior probabilities was gone. Inverse probability was stoutly defended as the only way to account for learning from experience, and priors the correct expression of individual belief or previous knowledge. The Simplicity Postulate reconciles the probability calculus with the behavior of physicists in the face of an infinite number of possible laws. Such laws are extracted from error-strewn measurements, yet the physicist's "predilection for the simple law is so strong that he will retain it, even when it does not fit the observations exactly, in spite of the existence

the tidal theory once occurred; and therefore that the tidal theory is true." Jeffreys 1924, p. 259. Jeffreys also used probability calculations in his book to argue that an extremely distended primitive sun, perhaps with a diameter that of the present orbit of Neptune, as in Jeans's version of the theory, was not required for the initial disruptive encounter posited by the tidal theory to be probable. See Jeffreys 1924, pp. 257–8.

[80] Jeffreys 1931, p. 7.

[81] Jeffreys 1931, p. 19. In *Scientific Inference*, Jeffreys introduced the now-standard notation $P(p|q)$ for the probability of a proposition p conditional on data q. In the papers with Wrinch he had used $P(p:q)$; Keynes and Johnson had used the McColl notation, p/q. (Jeffreys's notation is convenient for epistemic probability, but can be ambiguous when used for statistical hypotheses. See Barnard 1989, p. 260.)

of complex laws that do fit them exactly. Simplicity is a better guarantee of probability than accuracy of fit." Jeffreys was explicit that his theory of science was an attempt at a coherent description of practice. "By analyzing the processes involved in our forward scientific reasoning we detect the fundamental postulate that it is possible to learn from experience. This is a primitive postulate, presumably on the frontiers between *a priori* and empirical knowledge. The status of the laws of probability and the simplicity postulate is that of inferences from this principle."[82] The frequency theory, in contrast, depended on abstractions, and thus on operations that could not be carried out in practice. Infinite series and limiting values could only be the products of an inferential system, not the basis. Jeffreys regarded the frequency theory as exemplifying the mistake of working from axioms rather than backward from the raw data of sensations.

In the second part of *Scientific Inference*, Jeffreys showed how his model could account for the phenomenological development of real scientific theories. This was a program he had started with Wrinch during the relativity debate, and subsequently extended to geometry and mensuration. Jeffreys argued that the various concepts of dynamics – rigidity, displacement, space – were generalized in a non-quantitative sort of way from primary sensations of bodies and the comparison of lengths. From rigidity and the operation of ordering distances came a series of marks on a straight edge; thence a uniformly-graduated scale; thence – via a series of empirical postulates used to attach real numbers to pairs of points compared with the scale – a quantitative system of measurement. Jeffreys continued with a phenomenal description of the measurement of angles; trigonometric functions and the relations between them; abstractions of perpendicularity, the properties of triangles, and a set of Cartesian coordinates.[83]

Similar approaches had been taken by the positivists Mach and Kirchoff, and by Pearson in the *Grammar*. But Jeffreys's treatment was more mathematical and complete. He started, for example, with an operational account of the idea of number, of basic mathematics, and of physical magnitudes, using the idea of classes developed by Russell and Whitehead and extended by Wittgenstein and Ramsey. And having derived notions of length, time, and mass, he produced an account of Newton's laws of motion and of gravity, and

---

[82] Jeffreys 1931, pp. 47–8.

[83] The difference between Jeffreys's concepts and those usually meant by these words was in language: "It will be noticed that we have not defined the term 'plane' as such, but only the expressions 'lie in one plane,' 'in the same plane,' and 'the angle between the planes.' Thus we can attach meaning to these terms even though no physical plane has been constructed." Jeffreys 1931, p. 20.

finally – from a consideration of moving coordinate systems and the evidence of the Michelson–Morley experiment – of the principle, then the theory, of relativity.[84]

In each case, the procedure was the same: increasingly sophisticated concepts were abstracted as combinations of quantities conserved during the observed processes.[85] Velocity and acceleration came as derivatives of coordinates, for example, then followed kinetic energy, force, and angular momentum. Jeffreys went so far as to argue that this process described the historical development of science. Newton's laws of motion and gravity followed from observations of binary stars and the orbits of the moon; discrepancies in the motion of planets – for example, eccentricities, periodic disturbances, and secular progression – were incorporated as the effects of more distant bodies. Each time, the theory of probability was used to assess the claims of rival explanations. Could the gravitational effect of additional matter – say from the solar corona – within the orbit of Mercury be sufficient to account for the discrepancy of the perihelion shift within the usual Newtonian framework? No: estimated from the reflected light, the amount of any such matter is not enough for this explanation to be probable. Simplicity was also a factor. From generalized expansions of the components of Einstein's gravitational tensor, Jeffreys argued that there were no observationally superfluous terms in Einstein's expression. And since time was on a footing similar to the position coordinates, the equations of relativity could be reduced to a simpler form than those of Newtonian dynamics. For the moment, then, Einstein's theory has a high and climbing probability.[86]

---

[84] The Michelson–Morley experiment was in fact a series of precise experiments carried out by Albert Michelson, later with Edward Morley, in the 1880s. Using the interference properties of light, the experiment was intended to measure the speed of the Earth relative to the aether. The persistent null result, which seemed to lead to the counter-intuitive conclusion that the speed of light does not depend on the motion of its source, could later be explained under Einstein's special theory of relativity.

[85] Jeffreys noted that often these abstract concepts could also be associated with physical properties. For example, the various forms of heat and energy considered in thermodynamics could be related to the movements of constituent particles.

[86] The extent to which Jeffreys regarded this section of his book as a genuine history of science rather than a heuristic device is unclear. His account of the inference of the laws of motion from observations of the planets is tendentious in the extreme. Yet a descriptive reading would be in keeping with his positivism. Commenting that Einstein's gravitational theory was in fact obtained via general relativity, and not to accommodate wayward observations, such as the long-recognized anomalous perihelion of Mercury, Jeffreys noted, p. 186: "I think, however, that it is really rather in the nature of an accident that Einstein's law was obtained by his method and not by one very like that just given."

Pearson had written that the "unity of science consists alone in its method, not its material."[87] 'Science' was the process of coordinating and ordering the relationships between facts, and the difference between the traditional natural sciences and philosophy, or any other branch of inquiry, was simply of the sorts of facts under consideration. Jeffreys emphasized the point. The mensuration work had shown that there was no distinction between common-sense learning and science. ("Any result we offer must agree with common-sense and with results that can be logically or mathematically deduced from common-sense."[88]) The everyday examples of inference Jeffreys used in his book – such as catching a train – were not merely heuristic simplifications; all experience was to be assessed probabilistically.

Jeffreys's interest in psychoanalysis had not waned since the early twenties. In 1924 he had published a paper examining certain psychoanalytic themes of Ibsen's *Peer Gynt* in the *Psychoanalytic Review* (the joint editors of which, Smith Ely Jelliffe and William Alanson White, he had contacted while in America investigating the Harvard post). From early 1925 he formed part of a small Cambridge group that met occasionally to discuss psychoanalytic subjects.[89] In *Scientific Inference* he took the opportunity to extend and make explicit his earlier justification of the nascent science. Pearson had shown that learning, based on sensations, is at heart a matter for the individual. Jeffreys regarded the work of Freud – who, along with Karl Pearson and Sir George Darwin, constituted his private pantheon, and who he was to propose as a foreign member of the Royal Society in 1936 – on the analysis of dreams and the concept of the 'subconscious' as a case in point. In the same way that sensations of objects and events have been abstracted according to the probability calculus through notions of mass and force to the equations of dynamics, so does Freud form psychological concepts from a gradual refinement of the categories that summarize and coordinate sense-data. This is good

---

[87] Pearson [1911], p. 12.

[88] Wrinch and Jeffreys 1921a, p. 378.

[89] The group's members included Arthur Tansley, John Rickman, Lionel Penrose, James Strachey, and Frank Ramsey. (Ramsey died unexpectedly of a liver complaint in 1930, aged twenty-six, and though Jeffreys visited him in hospital during his illness, it was only after his death that Jeffreys discovered they had shared an interest in probability as well as psychoanalysis.) The first meeting was held in Jeffreys's rooms on 2 March 1925. Penrose reported on 'Psychoanalysis and Chess,' and Jeffreys on 'Psychoanalysis and Death Duties.' Strachey recalled this first meeting as 'gloomy,' and Jeffreys as "a rather rugged figure, like a wire haired terrier," who was "incredibly dull-minded, and comes so much from the North as to be almost incomprehensible" (Strachey 1986, p. 223). For more on the group, see Cameron and Forrester 2000.

science, not quackery. "It is precisely the utility of concepts in summarizing existing knowledge that makes it possible to keep scientific facts classified and accessible, and therefore to make progress as new laws are discovered." "At each stage the concept gets further away from the original facts; but at each stage also it makes it possible to infer more facts. The double aspect of the construction of concepts is not antagonistic to scientific method, but on the contrary is the very essence of it."[90] It was the critics of psychoanalysis, Jeffreys argued, who in refusing to take the subject seriously, or simply ignoring the data or the conclusions reasonably inferred from them, were being unscientific.[91]

Scientific Inference was more than a sanitized enlargement of the work with Wrinch. One novelty was Jeffreys's attempt at a quantified version of the Simplicity Postulate. Schrödinger's formulation of quantum mechanics suggested that perhaps the possible quantitative laws of science could, when suitably differentiated, be expressed as equations with numerical constants that were integers, rather than simply rational numbers as Jeffreys had assumed in 1923. A measure of a law's complexity could thus be taken as the sum of an equation's order, degree, and the absolute values of any coefficients. With the constraint that all prior probabilities must form a convergent series, the prior probability of any given law could be given numerically.[92]

Despite Jeffreys's efforts to render inverse probability in usable form, the reaction to his book was less than favorable, and sales disappointing.[93] Physicists were respectful but unenthusiastic, and ignored Jeffreys's discussion of relativity. For statisticians, the emphasis on the epistemic interpretation of probability was retrograde: they still used Bayes's theorem, but at least had the decency not to trumpet the fact. Like an alcoholic uncle, inverse probability could be turned to in emergencies but was an object of shame nonetheless. It was certainly no basis for a theory of inference. Philosophers too were dismissive. The exposition of probability was similar to that of Keynes, and did

---

[90] Jeffreys 1931, p. 196–7. Jeffreys was responding to critics of Freud, including his colleague Joseph Breuer, who accused him of presuming that the existence of a word meant that the concept it represents is real.

[91] Not only could the judgments of scientists be modeled with inverse probability, Jeffreys argued, so too could an individual's psychological development. He suggested that the subject be studied with the analysis of dreams, since these constitute the purest and most natural form of mental process, uncontaminated or constrained by conscious criticism. He predicted that even in dreams, the arrangement of mental processes would be shown to follow scientific laws, though these might be the crazy laws extrapolated by the child from insufficient and unrepresentative data. Jeffreys 1931, p. 204.

[92] Jeffreys 1931, p. 45.

[93] Jeffreys to Fisher, 5 June 1937 (Bennett 1990, p. 164).

nothing to explain why a logical relationship should have anything to do with a degree of belief. Moreover, in relegating his discussion of the frequency theory and competing accounts of scientific knowledge to the end of the book, Jeffreys was accused by reviewers of concealing the problematic nature of probability. Besides, they continued, his simple insistence that the success of science compelled acceptance of his theory was hardly persuasive for anyone with an alternative account. And his treatment of miscellaneous philosophical questions was a rag-bag. The most favorable notice was that of Ernest Jones, Jeffreys's former analyst, who approvingly quoted his defense of the methods of psychology and psychoanalysis at length in the *International Journal of Psycho-Analysis*.

4.11. JEFFREYS AND PRIOR PROBABILITIES

4.11.1. The status of prior probabilities

In *Scientific Inference* and Jeffreys's earlier papers with Wrinch, prior knowledge had a somewhat ambiguous standing. Officially, the position was clear enough: it was a fundamental postulate that human reasoning must accord with Bayesian probability theory in order to account for scientific inference, and, more generally, the process of learning from experience. Prior probabilities were part of this model of science, and any 'rules' governing their assignment – such as the Simplicity Postulate – were just formal ways to describe the behavior of scientists, and should be followed only to ensure consistency. This was a matter of description, not justification. "We have to decide for ourselves which results common-sense requires, then to consider out of which assumptions these can best be obtained, and then to see how far the assumptions themselves fit in with common-sense."[94]

Occasionally, however, the choice of prior probabilities seemed to take on a different guise. A particular form was preferred over others not merely because of its normative status as describing scientists' preferences, but because the success of scientific inquiry, and indeed of commonsense intuition, indicated that such a prior also reflected some contingent feature of the universe. This of course was Broad's conclusion – that the success of induction implied that linguistic categories were roughly congruent with the causal 'kinds' that constituted the universe – and since, as Jeffreys put it to Fisher, Broad's papers "influenced me to the extent that I assimilated the ideas and forgot where they came from," this dual aspect of prior probabilities is more evident in the earlier

---

[94] Wrinch and Jeffreys 1921a, p. 378.

work with Wrinch.[95] For instance, the paper of 1921 states that since some highly complex laws *are* known, such as that concerning the deformation of metals under stress, the prevalence of simple yet accurate laws, such as the Newtonian law of gravitation, is not merely a result of our investigation procedure, but is a quality of nature reflecting that "the simple law may be supposed to be *ipso facto* more probable than a complex one."[96] In the same way, Wrinch and Jeffreys's assumption that scientific laws be differentiable is sometimes glossed as a practical matter – necessary to remove continuously-variable constants so that simplicity can be compared – and sometimes as more an empirical fact or even logical necessity. Likewise, they are slippery over continua and enumerable sets. Does the statement that "continua have no place in physics," used to justify the quantification of simplicity, mean that continua have no place in the physical world, or in the *study* of the physical world?[97] Do the laws of physics form an enumerable set, or is it that we only admit those that do? The clearest contradiction with their declaration that the Simplicity Postulate cannot, and is not intended to, prove facts about the world is the 1921 statement that some laws can be dismissed as *a priori* highly unlikely, rather than as inconsistent with the process according to which all other scientific laws are inferred. An example is the suggested 2.000000016 gravitational power law, proposed to account for the anomalous perihelion of Mercury.[98]

This inconsistency persisted into *Scientific Inference*, and the exact relation between these interpretations of the prior probability remained unclear. In this sense, Jeffreys's theory of scientific inference shared a status with the eighteenth-century doctrine of chances. Both purported to be descriptive, but in describing the behavior of 'men of quality' both exerted a prescriptive force. In Jeffreys's case, the 'men of quality' were scientists, whose method was simply a sophisticated form of commonsense. The obligation on these scientists to adopt his assessments of prior distributions was to ensure consistency and uniformity with the collective approach to research. But consistency was only virtuous because of the success of scientific inquiry, and the choice of prior probabilities never simply a matter of consistency, since a probability,

[95] Jeffreys to Fisher, 1 March 1934 (Bennett 1990, p. 152).
[96] Wrinch and Jeffreys 1921a, p. 380.
[97] Wrinch and Jeffreys 1923b, p. 369. There is no immediate contradiction in Wrinch and Jeffreys's requirements here, since operations like differentiation, which seem based on continua, can be redefined for discrete functions.
[98] Wrinch and Jeffreys 1921a, p. 389. Jeffreys repeated the argument in a slightly different form in chapter 4 of *Scientific Inference*. Jeffreys 1931, pp. 50–1; see also p. 186.

by its nature, reflected individual belief. One effect of this was Jeffreys's desire to find accurate forms of prior probability, his rules for assigning *a priori* probabilities to laws, based on their simplicity, being a glaring example.[99] But Jeffreys sharpened his understanding of priors shortly after the publication of *Scientific Inference*, provoked in part by a paper by the biologist J.B.S. Haldane, and a subsequent response from Fisher.

4.11.2. J.B.S. Haldane's paper

Jeffreys was not the only scientist using inverse probability. Haldane, who interpreted probability in the clear epistemic sense as representing imperfect knowledge, had been considering the theory for problems in genetics, and he presented his conclusions to the Cambridge Philosophical Society in late 1931.[100] The same lottery-machine characteristics of genetics that invite a frequency interpretation of probability – ready interpretation as a population-based model, clearly-defined combinatorial outcomes, independent trials, long-run stability – make it also a close realization of the classical, balls-in-urn model of sampling. Haldane used the standard Laplacean analysis to update an empirical prior probability, concocted from data relating to genetic 'cross over' effects, with random sampling data. He pointed out an 'absurdity': due to the asymmetry in the prior, the most probable value – that is, the peak of the posterior distribution – will not in general coincide with the mean. For example, when drawing two samples of the same size from an urn of balls, the composition of the second sample should be expected to differ from that of the first.

Haldane's solution was to tailor the prior to specific circumstances. If we are dealing with a very small probability, for instance, the number of successes from even a large number of trials will be such that only an order of magnitude for this probability rather than its absolute value can be inferred with any confidence. The 8 anomalous alpha-ray tracks observed in a series of cloud chamber experiments by Blackett out of 415,000 cases give an idea of the scale of the effect, not an accurate assessment of the ratio in question. In such cases, it is natural to assume that the order of magnitude of the probability rather

---

[99] Jeffreys 1931, p. 46.

[100] Haldane 1932a. Haldane was a Marxist, and took the view that with full knowledge any probability, including those applying to moral decisions or human actions, could be replaced by a deterministic statement of near certainty. Though conceding that quantum mechanics might imply some slight indeterminacy in the world, he quantified this irreducible chance element as no greater than 0.05. See Haldane 1932b.

than the value of the probability itself is uniformly distributed. For the alpha-ray experiment, that would be to assume that the probability of an anomolous track is as likely to be around a thousandth of a percent as a hundreth or a ten-thousandth. This is equivalent to taking logarithmic values of the probability in the approximate neighborhood as initially equally probable – that is, assuming that the prior probability of the probability of an anomolous track, p, is not constant, but proportional to $1/p$ – and yields the required result when propagated through Bayes's theorem that the mean value of the posterior probability is equal to its most probable value.

Haldane's paper was spotted by Fisher. Though previously he had tended to restrict his criticisms of inverse probability to a statistical audience, Fisher was stung into action. Haldane's casual use of terminology, and conflation of epistemic and statistical probabilities,[101] reinforced Fisher's impression that the survival of inverse probability was more a consequence of insufficient schooling than a definite wish to advocate the epistemic interpretation. Even so, Haldane was not a reactionary biometrician, but a geneticist, working like Fisher to synthesize the Mendelian approach with Darwin's theory of natural selection. Fisher was anxious that this work be not infected by talk of prior probabilities. Further, Haldane had magnified his crime by declaring his "formal solution" to be a generalization and improvement of Fisher's supposedly approximate Method of Maximum Likelihood, and had accused Fisher of involving the Principle of Insufficient Reason in his theory as a "tacit assumption."[102] It was necessary to take a robust line.

Fisher decided to use the opportunity to hammer home his attack on inverse probability, and to advertise the alternative methods he had devised for problems of uncertain inference. He gave Haldane a tutorial. Inverse probability was inconsistent and without foundation. When we had precise knowledge of the sorts of classes we were dealing with, induction could be expressed with mathematical exactitude.[103] As in 1922, Fisher recast the problem of induction in frequency terms, and presented the eighteenth- and nineteenth-century pioneers as proto-frequentists. It was subsequently mathematicians – and recent

---

[101] See, for example, Haldane's discussion of reasons to believe that every even number greater than two can be represented as the sum of two primes.

[102] Haldane claimed his use of prior probabilities allowed "the deduction of Fisher's results without introducing concepts other than those found in the theory of direct probability." Haldane 1932a, p. 60.

[103] "[I]f we can assume that our unknown population has been chosen at random from a super-population, or population of populations, the characteristics of which are completely specified from a priori knowledge, then the statement of our inferences from the sample to the population can be put into a purely deductive form, and expressed in terms of mathematical probability." Fisher 1932, p. 257.

offenders such as Haldane – inexperienced in the processes of induction, who mistakenly applied the same procedure to cases in which our prior knowledge was not definite or specifiable. They were stumbling in dangerous territory. Vague knowledge is hardly an appropriate ingredient for precise mathematical inference. Certainly, the influence of the prior probability decreases with more information. But since in practice we always have finite quantities of data, it is never certain how dangerous erroneous priors will be. That a wrong procedure becomes an acceptable approximation in some cases is hardly a reason to use it in others. The occasional usefulness of mathematical probability for induction should not deceive: it is an inappropriate tool for uncertain inference. Instead, the method of maximum likelihood "possesses uniquely those sampling properties which are required of a satisfactory estimate."[104] Though not a probability, because not obeying the usual laws of probability, the likelihood sums up all the data, and is thus the only basis for purely inductive reasoning. It can serve – if you must – as good a measure of a "degree of rational belief" as a probability.[105]

Jeffreys took note of Haldane's work and Fisher's reply. Sampling was of course precisely the issue that in Broad's version had originally provoked his 1919 analysis with Wrinch of probabilistic induction. Yet with his more sophisticated model of scientific inference focused on quantitative laws and uncertain measurements, Jeffreys no longer found the issue as pressing. Haldane's genetics example really *was* like drawing balls from an urn. There was no reason to suspect a general law in his case; indeed, quite the reverse: genetic experiments were *characterized* by a random distribution of gametes. Hence: "There seems to be no reason to deny Laplace's hypothesis [that all ratios of balls in the urn are initially equally likely] where we have no previous relevant knowledge."[106] How then is the 'absurd' result of Haldane explained? Jeffreys thought that Haldane's intuition was, quite simply, mistaken. Since the possible number of black balls – if black is the color of

---

[104] Fisher 1932, p. 259.

[105] Fisher had no wish to humiliate Haldane further. Nevertheless, Haldane's suggestion that Fisher's work was an approximation to inverse probability, and could be better obtained by the introduction of prior probabilities, could not pass without comment. Fisher repeated that he had explicitly rejected priors, not used them 'tacitly,' and waspishly concluded that, "in so far as I have been guilty of a theory," its whole point and value is its independence from prior probabilities. (Fisher 1932, p. 261.) Haldane, however, was still confusing likelihood methods with the maximizing of Bayesian-type probability twenty-five years later. See Haldane 1957.

[106] Jeffreys 1932c, p. 86. While recognizing the assignment of prior probabilities to be "greatest stumbling block in the theory of probability," Jeffreys satisfied himself in *Scientific Inference* that in many circumstances Laplace's form accurately expressed a lack of knowledge. Jeffreys 1931, p. 20; see also pp. 24–35; pp. 191–197.

interest – in each sample is bounded by the size of the sample, we should *not* expect the fraction of black balls in a second sample to be the same as in the first. A small number of blacks in the first sample is more likely to lead to an underestimate than an overestimate of the actual ratio; after all, there could still be some black balls in the urn, and hence in a second sample, even if none of the first sample were black. Likewise, we should not predict that all the balls in the urn are black even if all the balls in the first sample are.[107] The issue of prior probabilities, however, was still of methodological importance. Jeffreys objected to Fisher's point that priors were irrelevant for induction. On the contrary, since the likelihood function merely summarized the sample, any inference concerning the whole class *required* some additional information.

Though sampling was no longer relevant for scientific laws, it was still to the point in Broad's original case of qualitative laws of the 'all crows are black' variety. Jeffreys had briefly tackled this issue in *Scientific Inference*, and, to account for the commonsense expectation that some general law is

---

[107] This example illustrates the general point that 'intuitions' are trained by experience. For Haldane, the urn modeled a genetics experiment that produced data in the form of new plant crosses. Hence his concern was that the *ratio predicted for the next sample* be intuitively plausible. For Jeffreys, the urn was if anything a crude model of a physical law that could be used to make general predictions. Hence his concern was not so much the next sample, but the inferred value of the *ratio in the whole class of events*. Haldane's reciprocal prior gives his commonsense result that the expected ratio in the second sample – the mean value of the posterior distribution – is equal to that obtained in the first. Since the posterior distribution is asymmetric, however, it gives a value for the expected ratio in the whole class – the peak value of the posterior distribution – that lies some distance away. Jeffreys's uniform prior gives the expected ratio for the whole class to be that of the initial sample, but gives the expected ratio in a second sample different from that in the first. Both Jeffreys and Haldane agreed that the mean of the posterior distribution generated with the first sample was to be understood as the expected ratio of black balls in the second sample, and that the mode gave the expected ratio in the entire class, but this too could be disputed. In his reply to Haldane, Fisher wrote critically of the tendency of many inverse probabilists to disguise the mode of the posterior distribution as "the most probable value." This terminology, he argued, was not justified; the mode is not strictly a probability, and has significance only because it coincides with the invariant optimum value given by the method of maximum likelihood. "Two wholly arbitrary elements in the process have in fact canceled each other out, the non-invariant process of taking the mode, and the arbitrary assumption," that the function specifying the super-population is constant. "[H]ad the inverse probability distribution any objective reality at all we should certainly, at least for a single parameter, have preferred to take the mean or the median value." M.S. Bartlett further pointed out that from the Fisherian perspective, not only is the optimum not necessarily the most probable value, but that the term 'most probable value' does not have a clear experimental significance. "If we already knew or suspected the true value . . . of a parameter before the sample was taken, we should be interested then not in the *optimum* value . . . as the most probable value, but in the chance that a discrepancy at least as great as that between the true value and the *optimum* value should have arisen" (Fisher 1930a, p. 531; Bartlett 1933, p. 531).

usually applicable, had anticipated Haldane with a reciprocal prior probability as a form of simplicity postulate for qualitative laws. Yet this form was not wholly satisfactory. As he had himself pointed out with Wrinch over a decade earlier, it had the defect that a general law was presumed true until an exception was found. And Jeffreys's discussion retained the old tension that while the prior was to be viewed merely as an expression of general opinion, the success of induction implied that it also reflected an empirical quality about the world, in this case that certain properties tend to come bundled together, and were accounted for by shared and stable linguistic categories.[108] Haldane demonstrated a fresh approach to priors. He did not agonize over the interpretation of his distributions, nor worry over their general validity. Instead, he treated each instance separately, and either worked backward to a prior probability from his intuitive expectations, or, in exceptional cases, constructed distributions that were workable but clearly artificial. He had noted, for example, that in some continuous cases, an empty first sample drawn from the urn – no black balls at all – could result in nasty infinities in the posterior distribution for the ratio of black balls in the urn. No problem: assign a finite value of probability to the ratio being exactly equal to zero, and spread the remainder of the probability evenly over all other possible values.

Haldane's synthetic analysis suggested to Jeffreys a pragmatic stance toward prior probabilities. More important than distributions that correctly described general scientific judgment – and perhaps also reflected the constitution of the world – was that inverse probability be used by scientists for problems of inference. Priors should be workable starting-points rather than accurate determinations appropriate in all cases. Haldane's paper also gave Jeffreys the opportunity to demonstrate his new flexible approach. One of Haldane's artificial prior probabilities, when put in symmetrical form, led to the same most probable value of posterior distribution as did the uniform prior of Laplace. The uniform prior, in other words, was not a unique way to recover an expected ratio equal to that of the value observed in the first sample. This result could be used to answer finally Broad's problem of the black crows. Instead of a uniform or reciprocal prior probability for the sampling ratio, Jeffreys recommended packing some finite value of probability, k, into each of the extreme values, 0 and 1, and distributing the rest evenly. He showed that with k independent of the size of the class (and non-zero), the posterior distribution, following repeated viewing of black crows, peaks at a value also independent of the size of the class. Thus with no observations to the contrary, the probability of a general qualitative law can approach unity long before

[108] Jeffreys 1931, pp. 191–7.

the total sample drawn becomes large compared to the whole class.[109] The magnitude of k reflects our confidence that a general law operates, and can be tailored to specific circumstances. For example, assume we are certain that some general law applies – that is, that either all or none of the sampled class has a given property. Then k is set to $\frac{1}{2}$, which packs all the prior probability into the values 0 and 1, and a single observation will serve to collapse the posterior distribution into a single value. Jeffreys suggested that when we suspect but are not certain of a law, a good starting point might be to take $k = \frac{1}{3}$.

4.12. JEFFREYS'S POSITION IN 1932

Jeffreys's theory of probability evolved through the 1920s as he gradually reconceived an inquiry begun in Wrinch's (and Russell's) style into an expression of his own experience of science. But this was of a particular sort of science. His research in theoretical geophysics and astrophysics, though wide ranging, formed a unified enterprise: to build an understanding, through study of the formation of the Earth and planets, of the origin of the solar system. Evidence from diverse fields was combined, reduced into mathematical relationships, then extrapolated to tell on conditions at the center of the Earth or millions of years ago. Perhaps in no other field were as many remarkable inferences drawn from so ambiguous and indirect data. (Though sparse compared with the social measures of the biometricians, Fisher's agricultural data was abundant and well-behaved compared with Jeffreys's earthquake data.) Necessarily, these inferences were tentative. They were advanced not with certainty but with degrees of confidence that were updated or modified to account for new information.[110]

As a theory of such inferences, inverse probability seemed to fit the bill. The epistemic interpretation allowed a direct assessment of hypotheses concerning determinate but unknown situations – such as the state of the Earth's core or the conditions at the time of the condensation of the solar system – and their revision in the light of new knowledge. It licensed the extrapolation, and measured the reliability, of quantitative laws. Indeed, there was an

[109] Since the probability distribution remains finite for every other value of the ratio, an exception to the general law – a white crow, say – will not lead to awkward consequences. Indeed, following a single contrary observation, the probability distribution reverts to the Laplacean form.

[110] Jeffreys was a keen student of detective stories – particularly those of Austin Freeman – and made copious notes on each character's alibis and motivations. Another instance of drawing inferences from incomplete and unreliable data!

analogy between inverse probability and what were called 'inverse' geophysical methods. "[T]he problem of the physics of the earth's interior is to make physical inferences over a range of depths of over 6000 km from data determined only for a range of 2 km at the outside."[111] In the same way that effects were inverted via Bayes's theorem to shed light on probable causes, so observations at the Earth's surface, such as seismic travel time estimates, were inverted to judge internal properties, such as wave velocity distributions and thus geological structure. A frequency definition, in contrast, was hardly appropriate. Restricted to repeated sampling from a well-behaved population, and largely reserved for data reduction, it could apply neither to the diverse pool of data Jeffreys drew upon nor directly to the sorts of questions he was attempting to address.

Additional features of Jeffreys's practice also fit with this model of inference. Both geophysics and astrophysics are based on observation rather than experiment. Fisher performed extensive plant- and animal-breeding experiments. Jeffreys, however, apart from an early botanical 'experiment' on the effect of water on certain plants, and a later attempt to investigate laminar fluid flow by dropping ink into the river Cam from a row boat, performed no experiments in his career. Fisher demanded the answers to exactly-posed questions; Jeffreys waited for nature to reveal itself. As a theorist, he worked alone rather than as part of a research group. His theory of probability likewise concerned the combination and weighing of data rather than its collection, and in the selection of prior probabilities accounted for the expertise and experience of the individual. Moreover, with no structural difference between data derived from observation and experiment, the line between science and commonsense learning was blurred.[112] Not restricted to experimental science, the probability calculus thus became for Jeffreys a model of the fundamental process of learning. This cohered with the operational philosophy he had developed from Pearson, and the associated idea of scientific laws as ever-improving probability distributions. Scientific practice – and hence a commonsense version of probability – comes first. It is formalized, not justified, by a theory of probabilistic inference. Thus was Jeffreys's specialized experience of science reinforced as a standard for inferential learning.

---

[111] Jeffreys 1924, p. 1. See also Bolt 1991, p. 17; Cook 1990, pp. 320, 326.

[112] "There is no methodological difference between the data obtained by experiment and observation." Wrinch and Jeffreys 1923c, p. 2. See also Jeffreys 1931, p. 208.

5

The Fisher–Jeffreys Exchange, 1932–1934

5.1. ERRORS OF OBSERVATION AND SEISMOLOGY

In the preceding two chapters I have described how the statistical methods of Harold Jeffreys and R.A. Fisher evolved from their respective disciplinary experiences in geophysics and genetics. During the twenties their methods exerted little mutual influence. The statisticians who comprised Fisher's audience were too aware of the shortcomings of inverse probability for him to provide anything more than general criticisms as he cleared the way for his likelihood and fiducial techniques; Jeffreys intended his theory primarily as an account of inference to rival that of the philosophers, and his objections to the frequency interpretation had little impact in statistical circles. Following the publication of *Scientific Inference*, however, Jeffreys started to apply inverse arguments directly to data analysis. Between 1932 and 1934, he and Fisher clashed over one of these applications. The exchange forced each to re-examine his methods.

The genesis of the debate can be traced to Jeffreys's involvement with seismology. Following his work with Wrinch on the Oppau explosion, Jeffreys had increasingly come to see earthquakes as an ideal way to probe the Earth's interior. The empirical situation, however, was somewhat akin to that Fisher had found on arrival at Rothamsted. Though abundant, the data was in too raw and unrefined a form to be of much use in geophysical inference. Turner's *International Seismological Survey* had published extensive quarterly records since 1918, but the rudimentary calculations of average earthquake travel times were thought to be out by up to 30 seconds.[1] Jeffreys was appalled in particular by the published estimates of uncertainty. Some were little better than guesswork. Seismologists widely used the method of least squares, but seemed unaware that the allied error distributions properly applied only when the number of observations was large. High-order polynomials were used to fit

---

[1] Lapwood 1982, p. 74.

data, with no thought of testing for significance.[2] For his early seismological work, Jeffreys had been forced to rely on a combination of data from these published tables and from direct correspondence with experimenters.[3] By the end of the twenties, however, he began to see that he had better deal with the matter of errors himself. The task was two-fold. First, the observations needed to be weighed for accuracy and reduced in the culinary sense to yield smooth curves of earthquake travel times against angular distances through the Earth's interior. Then a measure of the reliability of these relationships was required.

Jeffreys had discussed experimental errors at length in *Scientific Inference*. As a phenomenalist, he saw errors not as irritating trifles that obscured real, Queteletian quantities, but as the numerical discrepancies between raw observation and the supposedly true values adopted largely by convention from some inferred general law. The term 'error,' he argued, is a verbal matter. Observations can never be 'wrong': as recordings of sense data, they simply *are*. (It had been this picture of errors that had led him to the Simplicity Principle: a scientist's commitment to a simple linking relationship is strong enough to be embedded in the language; clearly, the simple law has some weightier status than merely a convenient approximation.[4]) Yet for any quantitative observation, some component of this discrepancy will be due to the combination of reading mistakes, the limits on precision of the measuring instrument, and the inevitable variation of experimental conditions as the data set is gathered. Thus a formal mathematical treatment of errors is still required to gauge the efficacy of adopted values as predictors of observed values.

Jeffreys devoted most of a chapter of *Scientific Inference* to this problem. Working from a precise error characteristic to the expected distribution of subsequent measurements is a straightforward application of direct probability theory. In practice, though, we come from the other direction, and seek a most probable 'true' value and precision from a series of measurements drawn from an unknown distribution. This is a job for inverse probability. Fortunately, the likely form of error distribution is often suggested by the experimental

---

[2] HJ (undated note [1978] to A. Zellner).

[3] Perry Byerly, the director of the Seismographic Station at Berkeley, met Jeffreys in 1929 at a seismological conference in Pasadena, and subsequently supplied him with much data (B. Jeffreys 1992, p. 306).

[4] Jeffreys regarded the observed values to be the most 'fundamental,' followed by the simple law and then the adopted values. "The observed values are found; they exist because they are measured; and there is nothing more to be said. A simple law is found to fit them approximately. This is a statement of fact. Then by a conventional process we find adopted values close to the observed values that fit the law exactly." Jeffreys 1931, p. 54; see also Jeffreys [1937], pp. 247–8.

conditions. Screw threads give rise to periodic errors, a coarse measuring system to a uniform systematic distribution. Jeffreys also considered errors due to a straightforward misreading of the instrument, and contributions that affect only some of the observations or are of only one sign. Even when the form of error is known, this work is not mathematically trivial. Take the normal law, for example, justified in situations in which the measurement is influenced by many similar and independent contributions, each small compared with the magnitude of the measured quantity (though as Jeffreys pointed out, invariably applied even in cases known to be dominated by a single cause).[5] The assumption of normality is not immediately useful, since it describes a distribution of known width and location parameters. In practice, not only do we not know what these values are, but our evaluation of them will be refined as the data accumulates. In consequence, the required quantity – the posterior distribution of the true value given the observations – will not be normally distributed; indeed, neither generally will be the total integrated prior probability for the distribution of the observations.[6]

As with all inverse probability calculations, some prior distribution is required for the width and location parameters of the assumed normal distribution. Jeffreys pondered the matter. We don't know much about the true value of the location parameter, x, perhaps only that it lies within a factor of ten, otherwise we would not be bothering with the measurements. Thus initially we can take its probability as constant, at least over the wide range indicated by our choice of instrument. What about the width of the final distribution, which Jeffreys expressed by the 'precision constant,' h?[7] We cannot, he argued, assume a prior probability that is uniform over all reasonable values. Rather, the distribution will depend on the circumstances. The normal error law is appropriate for both astronomical measurements and lengths measured by difference from a graduated scale, for example, but typically the precision of the latter depends on the step size of the scale, while the former is a function of the intrinsic scatter of the data, often due to atmospheric fluctuations. Nevertheless, we can produce a prior probability that although not tailored to

---

[5] Jeffreys's discussion of the shortcomings of Gauss's proof of the normal law is flawed. See Jeffreys 1933b, Bartlett 1933, Jeffreys [1937], pp. 262–3.

[6] "Similarly the posterior probabilities are not of the normal form even when the normal law holds. It is the *component* probability from each pair of values of [the width and location parameters] that is referred to when we speak of the normal law of error; any attempt to compound probabilities destroys the normal form." Jeffreys 1931, p. 72.

[7] Jeffreys's precision constant, or modulus of precision, h, is a measure of the precision of the measurements from the grouping of the values. It is related to the standard deviation: $1/h = \sqrt{2}\sigma$.

individual measurements, can be generally applied as a first approximation. Consider an abstract situation in which we receive measurements with no clue to the magnitude of their precision at all. First, the precision constant is restricted to positive values, and thus is not infinitely variable in the same way as in principle is the location parameter. Second, its prior probability must presumably share a form with that for the standard error – proportional to the reciprocal of the precision constant – since there is no reason to prefer one of these expressions for the width of the distribution over the other. To account for these conditions, Jeffreys proposed dh/h as the prior probability, and anticipated Haldane by interpreting this form as representing no knowledge of the *magnitude* of errors.[8] It is, in other words, an application of the Principle of Insufficient Reason to the order of magnitude rather than the absolute value of the precision constant.[9] Jeffreys showed that this prior probability seemed acceptable when propagated through Bayes's Theorem.

---

[8] It is difficult to give a convincing practical justification of this form, since Jeffreys's conditions describe a situation that occurs rarely, if ever, in everyday life. We invariably have some idea of the precision of a set of values, for we have some knowledge of either the measuring apparatus, or the sorts of processes required to generate the set. The following analogy, unless rendered obsolete by exponential increases in computing power, might help. Imagine we have programmed a fast computer to factorize an input integer. Our program is primitive and far from efficient, perhaps trying all multiplicative combinations of integers lower than the input, but the factors of a series of trial inputs – 15, 164, 2088, say – have appeared instantaneously on the screen. Now we type in a much greater number, the long string of an international telephone number or our social security number. What is the probability that the factors will appear on the screen in some small time interval? Intuitively, it will depend on the elapsed time. Say we start a clock when we hit the return key. The answer might flash up within the first second, or if not perhaps by the end of another. But as time goes on, it seems less likely that a further second will be enough to solve the problem. Faced with a blank screen after ten minutes, our confidence in the solution appearing in another second is far less than it was at the start. The fact is that we have little idea how long the calculation will take. Factorization is computationally intensive, but then ours is a powerful and expensive machine. The problem might take milliseconds, minutes, hours, or even decades. The form of prior reflecting equal probability for each of these possibilities – i.e., that the probability of a solution arriving between ten and eleven seconds is the same as that between ten and eleven hours, or even ten and eleven years – is dh/h. With this form the same probability is squeezed into the precision constant doubling from 1 to 2 as 10 to 20 or even 1000 to 2000. "This is equivalent to assuming that if $h_1/h_2 = h_3/h_4$, h is as likely to lie between h_1 and h_2 as between h_3 and h_4; this was thought to be the best way of expressing the condition that there is no previous knowledge of the magnitude of errors." Jeffreys 1932b, p. 48.

[9] I. J. Good has described the 'Jeffreys-Haldane' prior as the "first explicit use of the concept of invariance in statistics." Though like the uniform prior 'improper,' in the sense of integrating to infinity rather than one, the Jeffreys–Haldane prior can be approximated closely by a log-Cauchy distribution. This proper prior is useful in some cases, e.g., for a Bayes factor to represent weight of evidence. Good 1980, p. 25.

The posterior probabilities are in general entangled, and not cleanly separable into functions for x and h. Yet for large numbers of observations, the usual Gaussian distribution is recovered as an approximation.[10]

All of Jeffreys's mathematical techniques were developed for specific applications, and the new dh/h prior for the normal distribution was no exception.[11] It could be pressed into service immediately. Jeffreys had been considering the problem of outliers. Seismograms can be difficult to read. It is often difficult to tell when one wave ends and the next begins; the secondary wave, S, and a certain higher-order wave, termed SKS, overlap beyond 85°, for example. The possibility of misidentification mixes data from error distributions with differing degrees of scatter and possibly a systematic displacement.[12] A discrepant data-point can therefore never be assigned with certainty. It could be a chance outlier from the distribution of interest, or simply a misidentification. How then should such data be combined to yield reliable mean values?

This was similar to an old problem in astronomy. The scatter of astronomical observations has easy interpretation in terms of real and error components, and as described in Chapter 2, astronomers had from the seventeenth century led the way in finding probabilistic methods for the combination of data. Nevertheless, the correct attitude to the occasional appearance of very incongruous data was still a matter of dispute. Least squares soon gained acceptance as an objective method for data analysis, but the assessment of outliers seemed more a matter for the individual astronomer, with his expert feel for the reliability of data and intimate knowledge of the particular experimental conditions involved. Several probabilistic criteria had been proposed – Benjamin Peirce, astronomer father of the philosopher C.S. Peirce, published the first in 1852 – but eccentric data-points were usually rejected with an appeal to a rule of thumb. In 1863, William Chauvenet recommended ignoring observations beyond four standard deviations from the mean; fourteen years later, Wilhelm Jordan suggested this be restricted to three standard deviations.[13]

---

[10] This peaks for x at the mean value of the observations, with the usual standard deviation of $\sigma'/\sqrt{n}$, where σ' is the standard residual of the n observations; and for h very near $1/\sqrt{2}\sigma'$, with a standard deviation dropping as $1/(2\sigma'\sqrt{(n+1)})$. See Jeffreys 1931, pp. 66–70.

[11] The geophysicist E.R. Lapwood found no instance of Jeffreys's indulging in the reverse process, or casting around for an application only after developing a technique. Lapwood 1982, p. 74.

[12] Lapwood 1982, p. 76.

[13] Rules of thumb are still prevalent in data analysis. The historical fate of more objective measures has been variable. Generally, they have made more headway in larger research groups, in which the tasks of experimentation and analysis have been separated. See, e.g.,

Jeffreys had no quarrel with the arbitrariness of such rules of thumb, but worried about the effect of rejection on the new mean. Given that discrepant observations by definition have large standard deviations, their omission could shift the new mean sufficiently to require further rejections. In certain circumstances, the mean could crash around uncontrollably. A better solution would be a method of weighting; even the information of outliers should not be squandered. Jeffreys used dh/h to derive a weighting function to be used as an alternative to the rejection of outliers.[14] The dh/h prior also promised an alternative to the seismologists' sloppy use of least squares. Jeffreys used it to produce a probability distribution of error for the extracted coefficients that for few observations was wider than given by the standard expressions.[15] He sent his reconstruction of the theory of least squares to the Royal Society at the end of June 1932 for publication in the *Proceedings*.

5.2. FISHER RESPONDS

Fisher noticed the least squares paper and was irked. 'The Royal' was still giving Jeffreys space even though he showed no familiarity with recent developments in the subject of statistical analysis. This was especially irritating since Jeffreys had attended Fisher's demolition of Haldane's paper on inverse probability at the meeting of the Cambridge Philosophical Society only a couple of months previously. What to do? Fisher was more concerned with genetics than the less consequential geophysics, and had not previously troubled to engage with Jeffreys's thoughts on probabilistic induction, though he had read the early papers written with Wrinch.[16] This time, however, things were different. Jeffreys was applying his methods directly to the analysis of

---

Olesko 1995; Gigerenzer 1989, pp. 83–4. The rejection criteria mentioned in the text can be considered early significance tests, the distribution under scrutiny being normal.

[14] Jeffreys 1932a. This extended a discussion in *Scientific Inference*. According to Jeffreys's probabilistic rejection criterion, all data points within three standard deviations of the mean are retained, all further than six are rejected, and the cut-off for those in between depends on the number of data points. He concluded that the "common astronomical practice" of rejecting observations with residuals more than three times the probable error if they are otherwise suspect and five times regardless is too conservative, but that replacing the probable error with the standard error would lead to a better rule of thumb. Jeffreys 1931, pp. 80–3.

[15] Jeffreys 1932b. Jeffreys found that the posterior distribution for the least square coefficients was related to that previously obtained in *Scientific Inference* for the mean of several measurements of the same quantity. Although it yields the usual values of the most probable coefficients, the distribution is only normal as an approximation for many degrees of freedom, and becomes increasingly skewed for fewer data points. Jeffreys 1931, p. 69.

[16] Fisher 1922b, p. 46.

data, and though marginalized by the Cambridge statistical community, might still exert a disruptive influence on undergraduates if unchecked. After all, Haldane's paper showed that despite a general recognition that the assignment of prior probabilities by intuition or *a priori* was unsafe, even practicing scientists were still prepared to update past experimental or theoretical knowledge with the calculus of inverse probability. Moreover, Fisher regarded inverse probability to have "survived so long in spite of its unsatisfactory basis, because its critics have until recent times put forward nothing to replace it as a rational theory of learning by experience."[17] Yet a credible alternative was now available. Fiducial probability allowed objective probabilities to be attached directly to hypothetical statements. Jeffreys's continuing appeal to prior probabilities was not only improper but superfluous.

The statistical community was still dominated by biometricians, and remained both baffled by Fisher's intensely mathematical work, and unconvinced that the discipline needed such foundational reconstruction. Fiducial probability seemed to be making little headway. Nor was Fisher himself an accepted figure. The hostility that had greeted *Statistical Methods for Research Workers* in 1925 had softened little in the following years. Still lacking an academic position, Fisher was treated by the older generation of the Royal Statistical Society as an over-zealous upstart. They were not enamored of his blunt responses to criticism or perpetual sniping at Pearson. The influence of this energetic *enfant terrible*, grudgingly recognized, could be dismissed as exerted chiefly among vulgar American social scientists.[18] Hence the continuing need for Fisher to promote his body of statistical tools as more objective and coherent than the older, inverse methods. In this respect, Jeffreys's argument was particularly insidious. Not only was he claiming, in producing a unique prior probability distribution for an unknown parameter, to have sidestepped the most dependable of Fisher's standard objections to inverse probability, but the prior he had selected for the precision constant was fortuitously just that which, when combined with a uniform prior for the position parameter, resulted in a posterior probability numerically equivalent to the fiducial distribution. Jeffreys could no longer expect to peddle his theory without molestation. It was necessary to explode the inverse argument once and for all, and to re-emphasize fiducial

---

[17] Fisher 1930a, p. 531.

[18] See, for example, the discourteous reception of Fisher's paper "The Logic of Inductive Inference," read to the Royal Statistical Society in 1935, and Fisher's wryly bitter response, as discussed in chapter 6. Though his career academic output is roughly split evenly between genetics and statistics, Fisher was never employed in academe as a statistician.

probability as an objective and logically distinct alternative. Fisher responded immediately.

Fisher's rebuttal was received by the Royal Society in early November 1932. In claiming to have found the *a priori* distribution for the precision constant of a normally distributed variate, Jeffreys was "purporting to resolve in a particular case the primitive difficulty which besets all attempts to derive valid results of practical application from the theory of Inverse Probability." Fisher affected to be impressed: "That there should be a method of evolving such a piece of information by mathematical reasoning only, without recourse to observational material, would be in all respects remarkable, especially since the same principle of reasoning should, presumably, be applicable to obtain the distribution *a priori* of other statistical parameters."[19] But the surprise was short lived. Jeffreys had, quite simply, blundered – "What schoolboys would call a howler," Fisher wrote privately.[20]

Fisher focused his attack on a curious feature of Jeffreys's exposition. While working on the method of least squares, Jeffreys had come up with a ingenious new justification for his dh/h prior. Suppose we are told the values obtained from two measurements of some quantity, knowing nothing about the precision of these measurements save that the errors of observation are normally distributed. A third measurement is taken. What is the probability that this lies between the other two? Jeffreys's answer was unexpected: the probability is precisely one-third. "For the law says nothing about the order of occurrence of errors of different amounts, and therefore the middle one in magnitude is equally likely to be the first, second, or third made (provided, of course, that we know nothing about the probable range of error already)."[21] Jeffreys proceeded to equate this result with an expression for the posterior probability of the third measurement, then worked backward to find the consistent form of prior probability. His unique solution? The prior must be proportional to dh/h.

Fisher dismissed this as simply erroneous. Obviously, the probability of the third observation lying intermediate between the first two was 1/3 "if all three observations are made afresh for each test, but we may note at once that, for any particular population, the probability will generally be larger when the first two observations are far apart than when they are near together. This is important since, as will be seen, the fallacy of Jeffreys's argument consists just

19 Fisher 1933, p. 343.
20 Fisher to Bartlett, 26 September 1933 (Bennett 1990, p. 46).
21 Jeffreys 1932b, p. 49.

in assuming that the probability shall be 1/3, *independently of the distance apart of the first two observations*."[22] In fact, argued Fisher, the one-third property applies to *all* distributions; it could not possibly be used to produce an *a priori* distribution just for normal distributions. Indeed, one could easily construct a series of populations with some artificial distribution of h for which Jeffreys's claims would obviously break down. Fisher gave $df = ae^{-ah}dh$ as an example, for which the probability is $1/e$ that h exceeds $1/a$.

Fisher pinpointed the fallacy. The posterior probability of the third observation should be integrated not merely over all possible values of the unknown mean and precision constant of the distribution, as Jeffreys had done, but over all possible values of the initial observations too. Then, as expected, the equation would degenerate to an identity for all possible *a priori* distributions. Jeffreys's dh/h prior, together with a uniform distribution for x, generated a posterior numerically identical to the fiducial distribution for the precision constant. But this was purely a matter of luck, resulting from the fact that integrating over values of the precision constant with the substitution of Jeffreys's prior probability turned out to be equivalent to integrating over the initial observations. Hence the "deceptive plausibility" of Jeffrey's prior.[23] Yet a fiducial statement is logically distinct from a posterior distribution of inverse probability. Fisher explained the procedure. We can estimate the variance of a normal population from the sum of the squares of the deviations of a series of observations from their mean. The distribution of this estimate for random samples depends only on the width of the population sampled. Thus the distribution of the ratio of this estimate to the population width is independent of any *a priori* assumptions or unknown quantities, and can be calculated just from the number of observations. Indeed, this is a standard distribution and has been tabulated. Hence for a given probability we can state the value of the ratio that will be exceeded at that probability. This is an objective statement about frequencies in random trials, true whatever the actual distribution constant may turn out to be. Given this, we can easily work out a probability statement about the unknown parameter h in terms solely of known quantities. This fiducial statement remains true whatever might be the distribution of h *a priori*. Jeffreys's posterior, in contrast, refers only to a specified sub-set of possible samples.[24]

[22] Fisher 1933, p. 344.
[23] Fisher 1933, p. 348.
[24] See Fisher 1933, p. 347.

Fisher expected that his response to Jeffreys would end the matter. Haldane, after all, had been suitably chastened by expert intervention. But Jeffreys was not so easily cowed. Having already upheld inverse methods in his brief gloss on the Haldane–Fisher exchange read to the Cambridge Philosophical Society in November 1932, he followed Fisher's paper with a reiteration of the one-third calculation and a more detailed defense of his theory of probability in an article submitted to the Royal Society in February 1933. He also continued to defend the dh/h prior in public, discussing it at a meeting of the St. John's Mathematical Society in early 1933.[25]

Fisher deliberated. He had concentrated on what he thought to be a simple mistake in Jeffreys's mathematics, leaving more general criticisms of inverse probability either implicit or perfunctory. Clearly, though, Jeffreys's spirited stand called for a more complete debunking of the theory. All in good time. Fisher was immediately occupied with his career. In early 1933, Pearson had announced his impending retirement as Galton Professor of Eugenics at University College London, and the university authorities, in part encouraged by Haldane – himself recently appointed as Professor of Genetics, and eager to collaborate with Fisher – invited Fisher's application to succeed him. The problem was that Pearson's son, Egon, having worked in the department for some years, also had a strong claim on the post. After some wrangling, the post was split. Fisher was awarded the chair in eugenics, and Egon Pearson a position as reader heading a new department of statistics. From his arrival, Fisher's position at University College was uneasy. His staff were loyal to the departing Karl Pearson, and Egon, understandably protective of his new department, insisted that Fisher not give competing lectures on statistics.[26] But in December of 1933, Fisher found time to submit a combative reply to Jeffreys's second paper. The Mathematical Committee of the Royal Society, sensing that no resolution would be reached, wrote Fisher in February 1934 that a version of this contribution, together with a final rejoinder from Jeffreys of the same length, would be published on condition that each saw the other's article before acceptance. Fisher wrote Jeffreys to this effect on 23 February 1934. After a series of letters between the two men, Jeffreys submitted both papers to the Royal Society on 27 April 1934. They were printed together at the front of volume 146 of the *Proceedings*.

---

[25] George Barnard was in the audience. GAB (21.1.97).
[26] See Box 1978, pp. 257–9.

The dispute was partly a result of misunderstanding. Jeffreys tended to restrict his academic reading to the specific problems of interest at any time, and although he had requested direct from Fisher a copy of the great 1922 paper on statistics, had not subsequently kept up with developments in maximum likelihood and fiducial probability.[27] Believing that Fisher's methods were based on inverse probability, Jeffreys misinterpreted Fisher's paper as an attack on his least squares coefficients, which were similar in form to expressions previously obtained by Student.[28] This initial confusion was exacerbated by idiosyncrasies of exposition on both sides. Statisticians and geneticists alike found Fisher's work difficult. Possessed with a powerful insight, particularly in geometric as opposed to algebraic reasoning – typical is his early idea from statistical physics of using vectors in n-dimensional space to represent particular values of a parameter estimated from a sample of n measurements – Fisher was notorious in his published research both for failing to spell out reasoning that seemed to him obvious, and for occasionally providing flimsy arguments in place of rigorous, but forgotten, justifications arrived at by other means in his head.[29] While this intuitive approach was suitable for the intended readership of *Statistical Methods for Research Workers*, fellow statisticians complained that his terse expositions were mathematically dense, yet often rather less than rigorous. His bellicose response to such criticisms was in part the cause of his difficulty with the statistical establishment.

With Jeffreys the problem was different. Although he was inarticulate and halting in conversation, and universally agreed to be an horrendous lecturer, Jeffreys's writing, even relative to the high literary standards of the day, was exceptionally measured and clear. His later geophysical colleague Ken Bullen said he "understood Jeffreys best when he was in Cambridge and I was in Australia and we wrote postcards to each other," while the statistician George Barnard wrote that Jeffreys "combined being one of the best writers with being one of the worst lecturers I have ever known."[30] However, Jeffreys tended to assume that readers would be familiar with his previous work, and

---

[27] HJ (Jeffreys to Fisher, 9 June 1922); Jeffreys to Fisher, 1 March 1934 (Bennett 1990, p. 152). Fisher is mentioned only once, in a footnote, in *Scientific Inference*.

[28] HJ–DVL; Jeffreys [1961], p. 393.

[29] Fisher provided a geometrical argument for the one-third probability. (Fisher 1934, p. 2.) Yates and Mather suggest that this ability to tackle mathematical problems mentally – often via geometrical analogies or models – was a legacy of the paperless private instruction he had received as a boy on account of his poor eyesight. Yates and Mather 1963, p. 91.

[30] Bullen quoted in Lapwood 1982, p. 82; GAB. Alan Cook recalls the attendance at a 1946 course of Jeffreys's lectures dwindling to the point that only embarrassment at further halving the audience kept him there, and records a similar story told by D.J. Finney of the period 1937–8. Cook 1990, pp. 307–8.

in consequence, although his books are argued fully, the research papers are often scantily introduced, better understood as works in progress that reflect the evolution of his ideas and preoccupations. In part this was due to the sheer number of papers he wrote; his unfashionable subject matter, solitary work habits, and concentration on matters philosophical also served to isolate him from the sort of critical feedback that would have provoked a fuller explication.

Yet the confusion underlying the Jeffreys–Fisher dispute was more than stylistic. Fisher refused to address the issue on Jeffreys's terms. Though he knew Jeffreys took a probability to represent a degree of belief, Fisher's commitment to his own interpretation was such that he insisted on translating Jeffreys's measurement question into the language of his statistical model: "What distribution *a priori* should be assumed for the value of h, regarding it as a variate varying from population to population of the *ensemble* of populations which might have been sampled?"[31] Usually, the numerical equivalence of results obscured the distinctions between the mathematical – as opposed to philosophical – interpretations of the two forms of probability. Jeffreys, however, had unwittingly asked a question not directly intelligible in terms of an ensemble of populations. It was not a natural question for a frequentist to ask, and in recasting the problem as a matter of sampling from an infinite hypothetical population, Fisher mangled it.

5.4. THE MATHEMATICS OF THE DISPUTE

The mathematics of the dispute was cleared up by M.S. Bartlett, later a distinguished statistician, then a 22-year-old in his fourth, graduate year at Queens' College, Cambridge. Bartlett had read the first papers by Fisher and Jeffreys and was encouraged to contribute to the debate by John Wishart, a University Reader with responsibilities in both mathematics and agriculture. Wishart had the previous year started a new lecture course on statistics for the mathematicians, and though of the new statistical school – he had spent a year at Rothamsted with Fisher – got on with Jeffreys well enough to have offered preliminary comments on his statistical work. Yet Wishart had been aggrieved when Jeffreys had not only scheduled a rival course of lectures on probability, but had chosen the hour immediately before Wishart's on statistics.[32] He passed Bartlett's paper in March 1933 to Yule – who, still sprightly two years previously at sixty, had obtained a pilot's license – and Yule in turn

---

[31] Fisher 1933, p. 343. See too Lane 1980, p. 152.

[32] This presented a Hobson's choice for undergraduates: Wishart's lectures were notoriously as boring as Jeffreys's. IJG, GAB, MSB.

communicated it to the Royal Society.[33] It appeared in volume 141 of the *Proceedings*.[34]

Jeffreys's second paper had appeared by the time Bartlett's was published, and not wishing to intrude further into what was becoming a personal battle, he stepped quietly back. Yet Bartlett had identified a main source of contention. He had recently read the posthumously-published *The Foundations of Mathematics and Other Logical Essays* by his one-time lecturer in mathematical analysis, the brilliant young mathematician Frank Ramsey, and saw, through Ramsey's essay "Truth and Probability," that Fisher and Jeffreys were quite simply talking of different things.[35] Jeffreys's probability described a mental state, a degree of belief in a proposition relative to specified data; Fisher's was an objective quantity, uniquely determined by some natural quality, such as the construction of a gaming machine or the distribution of gametes in a plant, and had a value independent of any knowledge or background data. Bartlett followed Ramsey – and, of course, Poisson – in reserving the term 'probability' for Jeffreys's interpretation and 'chance' for Fisher's.[36] Probabilities clearly have little to do with frequencies. Chances, however, can be equated with frequencies in idealized infinite populations, and sets of observations regarded as random samples from such populations. This is equivalent to assuming that each observation obeys some definite and unchanging law of chance. Thus Fisher, from an exactly specified if numerically unknown distribution, had produced an expression for the chance of a given set of observations. His calculation was direct, in terms of the parameters of the particular population. Jeffreys, in contrast, had calculated the probability of the observations without any assumption concerning their distribution of chance. His distribution of prior probability was not, therefore, a statement of chance – referring in this particular case to the distribution of measuring conditions in the world – but instead encoded a state of knowledge analogous to that assumed by Fisher in his specification of chance distribution. In short, Jeffreys and Fisher were basing different sorts of statement on different background assumptions. No wonder they did not agree.

---

[33] Yule had been appointed a Cambridge lecturer in statistics in 1912, and worked in economics and agriculture. From 1913 he was a fellow at Jeffreys's college of St. John's, and occasionally lent Jeffreys his powerful Brunsviga calculating machine. Yule suffered with heart problems in 1931, and was henceforward a "semi-invalid." See Kendall 1952.

[34] Bartlett 1933. Lane 1980 made many of the same points in his gloss on the exchange.

[35] See Olkin 1989, p. 162; also commentary to Bartlett's paper in Bartlett 1989, vol. 1.

[36] Broad offered the definition: "An event may be said to be a matter of chance when no increase in our knowledge of the laws of nature, and no practicable increase in our knowledge of the facts that are connected with it, will appreciably alter its probability as compared with that of its alternatives." Broad 1922, p. 81.

As Bartlett explained, the two probability statements diverged when Fisher couched the question in terms of frequencies. Fisher had noted that the probability of the last of three independent observations taken from the same population lying between the first two is obviously $\frac{1}{3}$ "if all three observations are made afresh for each test." But in commenting that "for any particular population, the probability will generally be larger when the first two observations are far apart than when they are near together," he revealed a misunderstanding of the original question. Jeffreys did not consider any particular population. He assumed neither that the probability statement was conditional on fixed values of the mean and standard deviation, nor that it was conditional on fixed values of the mean, standard deviation, and particular values of the first two observations – two distinct statements Fisher thought Jeffreys had confused. Instead, Jeffreys's probability statement was relative to particular values for the observations, but unknown values for the mean and standard deviation. In this case, without knowledge of the standard deviation, it is not possible to say whether two observations are near or far apart, whatever might be their particular values.

The differences between these statements are best illustrated symbolically.[37] Fisher thought Jeffreys was equating the probability in a particular population,

$$P[\min (M_1, M_2) \leq M_3 \leq \max (M_1, M_2)|x, h]$$

that is, the probability that the third of the three measurements, M_1, M_2, and M_3, lies between the first two, given the values of the mean, x, and precision constant, h, of the normal distribution from which they are drawn, with

$$P[\min (M_1, M_2) \leq M_3 \leq \max (M_1, M_2)|M_1 = m_1, M_2 = m_2, x, h]$$

that is, the probability that the third of the three measurements lies between the first two, given the values of the mean and standard deviation of the normal distribution from which they are drawn, and the particular values of the first two observations.

The first is generally equal to $\frac{1}{3}$. As Fisher noted, this is a property of all distributions, and hence can reveal nothing about the distribution of h in normal distributions. The second will not generally equal $\frac{1}{3}$. Its value will depend on how close m_1 is to m_2 relative to the width of the distribution, h. The two statements cannot be equated unless the second is averaged over all pairs of initial observations.

---

[37] See Bartlett 1933, Lane 1980. Mine is similar to Lane's notation.

In fact, Jeffreys took:

$$P[\min (M_1, M_2) \leq M_3 \leq \max (M_1, M_2) | M_1 = m_1, M_2 = m_2]$$
$$= 1/3, \forall m_1, m_2$$

This can be read as the statement that the probability of the last of three independent observations taken from the same population lying between the first two, given the first two, is equal to one-third, whatever might be the specific values of the first two measurements involved.

Fisher's confusion is understandable. The $\frac{1}{3}$ business seems odd. The numerical values of the three hypothetical measurements Jeffreys considers give no indication of the precision with which they are measured. Even so, at first blush one might expect the probability to depend on the particular measurements: if the first two are wide apart, the third is more likely to separate them than if they are closely spaced. Yet Jeffreys's point is that since we have no knowledge of the precision of the measurements or the experimental circumstances, we cannot know what 'close together' means until more measurements are taken. Length measurements of 1.000 and 1.001 meters will not seem close if it turns out we are measuring the length of an optical cavity using an laser interferometer rather than a table with a tape-measure. Jeffreys's dh/h prior is equivalent to taking the logarithm of h to be uniformly distributed; in other words, to taking all *magnitudes* of the precision to be equally likely.

Fisher's reading of Jeffreys's question highlighted a major difference between the frequency and epistemic interpretations of probability. Prior probabilities in each case represented different sorts of statement. Fisher's probabilities were statements about frequencies existing in the world. He assumed that in claiming that the one-third argument was sound, Jeffreys must be referring to drawing a third measurement from a random population; his dh/h prior therefore must supposedly represent the distribution of normal populations, and hence measurement procedures, that existed in the world. But for Jeffreys a probability was not a known frequency, but a state of knowledge. ("By 'probability' I mean probability and not frequency, as Fisher seems to think, seeing that he introduces the latter word in restating my argument."[38]) His prior probability encoded not the distribution of actual measuring procedures, but the information specific to a particular one. Hence Fisher's sardonic opening comments – "That there should be a method of evolving such a piece of information by mathematical reasoning only, without recourse to observational

[38] Jeffreys 1933a, p. 523.

material, would be in all respects remarkable . . . " – seemed beside the point to Jeffreys, who readily agreed that his prior could not possibly represent a known frequency: "A prior probability is not a statement about frequency of occurrence in the world or any portion of it. Such a statement, if it can be established at all, must involve experience."[39] Prior probabilities are not opinions on the ratio of cases in the world. "My attitude, on the contrary, is that it is only on my view that it is possible to make any progress without expressing such an opinion . . . "

Consequently, the artificial distribution of h that Fisher offered as a counter-example missed the point of Jeffreys's formulation. Fisher's frequency element $df = ae^{-ah}dh$ represents knowledge of the sampling probability of h – specifically, that the probability is $1/e$ that h exceeds $1/a$ – whereas Jeffreys's assumption that we do not know which distribution we are dealing with is precisely the information that his prior probability is supposed to encode. "With such previous knowledge there would, of course, be no need for any further discussion of the form of $f(h)$; if it is already known, it is known, and there is no more to be said."[40]

5.5. PROBABILITY AND SCIENCE

Bartlett's gloss, that the confusion over the one-third argument was due to different definitions and background assumptions, was broadly accepted by both men. "I don't think there is much to add to Bartlett's discussion," wrote Jeffreys to Fisher, who conceded that it was "thoughtful."[41] Yet the distinction between probability and chance extended beyond a mere difference of statement. For both Fisher and Jeffreys, the concept of probability was intimately tied to the nature of scientific inquiry. Their incompatible views on the sorts of question a scientist could meaningfully ask infused not only their interpretations of probability, but the rest of their statistical and inferential apparatus too. Bartlett had cleared up the mathematics of the matter, but he had not shown how these differing views colored each associated theoretical framework. Nor had he provided a clear indication of the differing scientific methods of each man, or a convincing argument favoring one of these over the other. Hence there was to be no resolution. Fisher and Jeffreys continued their argument largely at cross purposes, not understanding that each was

---

[39] Jeffreys 1934, p. 12.
[40] Jeffreys 1933a, p. 531.
[41] Jeffreys to Fisher, 1 March 1934 (Bennett 1990, p. 155); Jeffreys 1934, p. 14; Fisher 1934, p. 7.

advocating not so much an alternative grounding for the calculus of probabilities as an entirely distinct approach to the scientific enterprise.

5.5.1. The status of prior probabilities

Associated with the pervasive difference over the *interpretation* of prior probabilities was a confusion over their *status*. Fisher's priors were just objective statements about frequencies. Jeffreys understanding of epistemic priors, though, was more nuanced, as his response to Haldane illustrates. In the exchange with Fisher, this came out in the crucial distinction between the status of prior and *a priori* probability distributions (terms not distinguished by Haldane). These are not comparable in the frequency interpretation. For Fisher, a probability was a frequency ratio in the world, to be assessed with knowledge. Thus prior probabilities – frequency ratios based on past experience – could be evaluated, while *a priori* probabilities could by definition not. In Jeffreys's epistemic interpretation, however, the two could usefully be distinguished. The solution to Broad's problem is an example of a prior distribution. It is not supposed to be a unique distribution, nor to represent a real state of knowledge: Jeffreys is quite frank that the variable k may be chosen for convenience or by convention. Instead, it is a rough way to encode previous knowledge or information, to be used merely as a convenient starting point for calculation. "The function of the prior probability is to state the alternatives to be tested in such a way that experience will be able to decide between them."[42] An *a priori* distribution, in contrast, was a unique measure constructed independent of experience.[43] Fisher would deny such a notion; Keynes too was leery of it. But Jeffreys used Bayes's theorem to show that on the epistemic view, "if we can assign a meaning to a probability on experience, we can certainly assign one without it."[44] Just as the theorem governs the updating of belief with new data, so it could be run backward to strip background information successively from a probability distribution.

Jeffreys's form of words here is ambiguous. Surely an assignment of probability on no experience is contrary to his notion of a probability as a *relation*

---

[42] Jeffreys 1934, p. 12 (see too Jeffreys 1931, p. 252). Jeffreys remarked on the previous page that: "Prior probabilities could logically be assigned in any way; they must in practice be assigned to correspond as closely as possible to our actual state of knowledge, and in such a way that the sort of general laws that we, in fact, consider capable of being established can acquire high probabilities as a result of sufficient experimental verification."

[43] Jeffreys 1933a, p. 524. See also Jeffreys 1931, p. 2: *a priori* knowledge is "applicable to the study of experience and not itself derived from experience."

[44] Jeffreys 1933a, p. 529.

between two propositions, one of which is a body of evidence or data? Yet when Jeffreys speaks of *a priori* probabilities, he means probabilities based not on no data, but on *a priori* data, which is that known independently of experience. Hence for Jeffreys an *a priori* probability does exist. "It is essential to the theory that a true *a priori* distribution of probability exists, that is, a probability assignable on our *a priori* data alone. That is to be distinguished from the prior probability, which is the probability before some particular test, and may rest on a certain amount of previous observational evidence."[45] The key idea is 'ignorance.' Fisher took this as a sort of background assumption that signified in this specific case a lack of information concerning the population of all possible measuring conditions. Jeffreys's claim to supply an *a priori* distribution from ignorance was therefore a contradiction. But Jeffreys considered 'ignorance' to refer to a *specific* state of knowledge – the *a priori* condition. "Complete ignorance *is* a state of knowledge, just as a statement that a vessel is empty is a statement of how much there is in it."[46] His prior probability was to be a formal expression of that knowledge, numerically precise as was the state of ignorance he was aiming to describe.

In practice, the difference between Jeffreys's prior and *a priori* probabilities was one of approach. Prior probabilities are usually not worth assessing precisely, because the relevant background information is vague and difficult to quantify. This difficulty does not mean that priors can simply be ignored. However, as Haldane had shown, a casual attitude can be taken toward them, and the true prior distribution approximated by some rough workaday function. ("When previous knowledge is complex the difficulty of evaluating a prior probability is one of sheer labour; nevertheless, I think that in many

---

[45] Jeffreys 1933a, p. 530. Shortly after the exchange with Fisher, Jeffreys used this distinction to recant the view, implicit in his discussion of the 2.000000016 power law in the papers with Wrinch, and later expanded in *Scientific Inference*, that *a priori* probabilities could be assigned to laws based on their simplicity: "It appears that at the present stage of scientific knowledge the prior probability of a simple quantitative law or a general one is not assessed *a priori* (that is, independently of experience)...It is inferred from the frequency of verification of such rules in the past...[T]he suggestion on p. 46 of my book that it is worth while to determine the number of possible quantitative laws of complexity, as there defined, less than a certain value, and to state their prior probabilities accordingly, ceases to have much interest." The Simplicity Postulate was properly a rule for the consistent assignment of prior probabilities. It was not an *a priori* but a prior principle, based on previous experience of the prevalence of simple laws. Jeffreys 1936, pp. 344–5, 346. When *Scientific Inference* was reissued in 1937, Jeffreys took the opportunity to reiterate his retraction in a new addenda. Jeffreys [1937], pp. 244–5, 249–50, 262.
[46] Jeffreys 1931, p. 20.

cases approximations can be obtained."[47]) *A priori* distributions are a different matter. Since the background knowledge is precisely specifiable, they can in principle be ascertained exactly. The issue in any particular circumstance is whether the relevant *a priori* conditions can be quantified, and if so, whether our supplementary experience is really so nearly negligible to make the *a priori* prior probability useful.

In the general case of measurements from a normal distribution, Jeffreys regarded the latter condition as satisfied. In practice we will always have some relevant information about the magnitude of the precision, yet since this information will often be marginal, the *a priori* distribution can serve as a good standard starting point. As for the former question, Jeffreys's answer had been that the state of 'ignorance' concerning the unknown precision can indeed be expressed exactly: we have just two measurements and the knowledge that they are from a normal distribution of unknown mean and standard deviation. How could this exact state of ignorance be quantified? Jeffreys pointed out that since starting knowledge need not necessarily be 'prior' in the temporal sense – it is merely one side in a mathematical equation – there are two distinct ways. The first is simply to consider what sort of distribution is consistent with our *a priori* knowledge; this was the route Jeffreys had followed in *Scientific Inference* when he produced the prior required to express an equal distribution of scale parameter. The second is to work Bayes's equation backwards, and to "investigate what distribution is consistent with facts otherwise known about the posterior probability on certain types of data"; this was the one-third argument he aired in the 1932 paper in the *Proceedings* of the Royal Society.[48] In both cases, dh/h was the result.

Jeffreys's particular reading of the status and role of *a priori* probability distributions is illustrated by his response to Fisher's accusation that the dh/h prior is 'improper.' As far as Fisher was concerned, for an *a priori* distribution to represent anything at all, it must be of a valid form. Yet Jeffreys's probability was an impossible distribution of populations. Probability distributions must integrate to unity – after all, h must have *some* value – but Jeffreys's dh/h diverges at both limits of the precision constant. Hence Fisher seized on Jeffreys's offhand comment that the "relation must break down for very small h, comparable with the reciprocal of the whole length of the scale used, and for large h comparable with the reciprocal of the step of the scale; but for the

[47] Jeffreys 1934, p. 13.
[48] Jeffreys 1933a, p. 531. Keynes had done something similar in working backward from intuitively desirable values of the arithmetic, geometric, harmonic, and other types of mean to the error distribution necessary to generate them as most probable values.

range of practically admissible values it appeared to be the most plausible distribution." This was manifestly an attempt at evasion.[49] "Jeffreys himself seems to feel some doubt as to the general validity of the distribution," he wrote. "It is, moreover, as Jeffreys, by his references to large and small values of h, clearly perceives, an impossible distribution *a priori*, since it gives a zero probability *a priori* for h lying between any finite limits, however far apart."[50]

Yet Jeffreys's blithe manner followed rather his having already discussed the matter in *Scientific Inference*. He replied to Fisher that since the distribution expressed a state of partial ignorance its impropriety was to be expected. Indeed, the divergence was a *necessary* feature of the prior probability distribution. Only when the distribution diverges is the integral indeterminate, consistent with having no previous knowledge relevant to the value of h. Moreover, if the form of prior probability involved any other power of h – Jeffreys assumed a power relationship from a consideration of dimensions – then the ratio of the total probability of h lying above, to that below, any specified value could be calculated. This would lead to definite conclusions about h, contrary to the assumption of ignorance. *"The fact is that my distribution is the only distribution of prior probability that is consistent with complete ignorance of the value of h."*[51] Further, he argued, this form has the desirable feature that after a single observation, the posterior distribution remains as dh/h. "This is what we should expect, since a single observation can tell us nothing about its own precision. It is only when we have two observations that a definite standard of precision, given by their separation, is introduced; and it is precisely then that the theory for the first time leads to definite results as to the probability that h exceeds a certain value, the integrals . . . converging at both limits. At all points the theory gives results in accordance with experience."[52] Nor was the divergence a practical problem. The choice of measuring apparatus, of certain range and scale characteristics, constitutes relevant background knowledge that rules out the extreme small and large regions of the distribution and thus serves to confine the measurement to a well-behaved area.[53]

---

[49] Jeffreys 1932b, p. 48.
[50] Fisher 1933, pp. 344, 348.
[51] Jeffreys 1933a, p. 531.
[52] Jeffreys 1933a, p. 532.
[53] "We choose the length of our scale so that all the measurements will be included within it easily; that is, the important values of h are large compared with the reciprocal of the length of the scale. Again, if the scatter of observations is comparable with the step of the scale, the finiteness of the step is a dominant source of error and the normal law does not apply at all." Jeffreys 1931, p. 67.

The status of priors provided a common thread of misunderstanding through-out the exchange. It emerged again during the discussion of the Principle of Insufficient Reason. From Fisher's prescriptive point of view, the inconsistency of this principle remained crippling to the theory of inverse probability. He drew attention to Jeffreys's claim that a probability was an undefined 'degree of belief,' unrelated to a frequency. Since "[o]bviously no mathematical theory can really be based on such verbal statements,"[54] Jeffreys must be using the Principle of Insufficient Reason as the sole basis for assigning numerical probabilities. Yet a probability is an objective measure, while the Principle of Insufficient Reason is not only subjective and impossible to verify experimentally, but arbitrary and inconsistent too. The principle was clearly too flimsy a base for the entire weight of a numerical theory of probability. Jeffreys's interpretation, therefore, was not viable.

But Jeffreys had preempted Fisher's criticism. He had already introduced 'the Principle of Non-sufficient Reason' in his second paper, and readily admitted that while Bayes's theorem provided the framework for probability, the principle was often used in practice to input numbers. He was also frank in drawing attention to the problematic nature of the rule. "This principle . . . seems to me so obvious as to hardly require statement; but it is at this point that many writers, having proceeded correctly so far, refuse to take the necessary further step and render the theory sterile or confused." He produced a case in point from Keynes: "If, to take an example, we have no information whatever as to the area or population of the countries of the world, a man is as likely to be an inhabitant of Ireland as of France. And on the same principle he is as likely to be an inhabitant of the British Isles as of France. And yet these conclusions are plainly inconsistent. For our first two propositions together yield the conclusion that he is twice as likely to be an inhabitant of the British Isles as of France."[55]

As discussed in Chapter 4, Keynes was an adherent of the degree-of-belief interpretation, but a harsh critic of the Principle of Insufficient Reason. Thus he had commented on the last paragraph: "Unless we argue, as I do not think we can, that the knowledge that the British Isles are composed of Great Britain and Ireland is a ground for supposing that a man is more likely to inhabit them than France, there is no way out of the contradiction. It is not plausible to maintain, when we are considering the relative populations of

[54] Fisher 1934, p. 2.
[55] Jeffreys 1933a, pp. 528–9.

different areas, that the number of names of subdivisions which are within our knowledge is, in the absence of any evidence as to their size, a piece of relevant evidence."[56]

Jeffreys was generally appreciative of Keynes's contributions to probability theory. But he regarded this as fallacious. The example of the countries was similar to the case of the colored book he had discussed in his review of the *Treatise*: the inconsistency is only apparent, and results from the assumption of different background data. The information in the word 'country' varies between the two cases. If some remote foreigner took Great Britain, Ireland, and France as three different countries, "he must assess their probabilities equally, and the probability that [the inhabitant] comes from Great Britain or Ireland *is* twice the probability that he comes from France. If, on the other hand, he considers that the British Isles are one country, of which Great Britain and Ireland are divisions, he must assign to the British Isles and France the same probability, dividing that assigned to the British Isles equally between Great Britain and Ireland. Keynes's dilemma does not exist, and is merely an indication of incomplete analysis of the nature of the data."[57]

Unlike Keynes, Jeffreys did not think that the Principle of Insufficient Reason needed to be defused. It was merely a starting point for calculation, intended primarily to express the possible options with as little prejudice as possible, and to not rule out any new possibility that subsequent evidence might suggest.[58] Therefore he was unimpressed with Keynes's criticism that a unique formulation of the principle could not be produced. One should not expect this. Instead, external considerations in each particular situation could be used to select the relevant equally-possible options, subject to the condition that the probability calculus must then capture the process of learning by experience. Then the principle could be applied perfectly consistently. Jeffreys's solution to Broad's problem was an example. The inference of qualitative, as opposed to quantitative, laws could indeed be modeled by the process of drawing balls from an urn. But we do not consider the world as a random collection of balls, we instead assume some sort of order. Jeffreys's recommendation that in many cases k should be set to $\frac{1}{3}$ could be taken as an application of the Principle of Insufficient Reason, not to all proportions of

[56] Keynes 1921, p. 44. Quoted in Jeffreys 1933a, pp. 528–9.
[57] Jeffreys 1933a, p. 529.
[58] "The principle of non-sufficient reason is intended to serve simply as an expression of lack of prejudice; in a sampling problem we want to give all constitutions of the whole class an equal chance of acquiring a high probability by experiment." Jeffreys to Fisher, 1 March 1934 (Bennett 1990, p. 154). See also Jeffreys 1931, pp. 21–2, 34–5.

balls in the urn, but to each of the three logically parallel statements that "all of the class is of a particular type," "none of the class are," and "some are."[59]

Fisher was puzzled; this was barely comprehensible as a defense. A probability had a unique and particular value, independent of the form of words used to describe it. Yet as Keynes had shown, the Principle of Insufficient Reason led to inconsistent probability assignments. Jeffreys did not seem to recognize the force of the objection that he had himself cited. His defense was purely a verbal dodge, already dealt with by Keynes. Of course the word 'country,' like many terms, can be employed in more than one sense. But preferences between these should be irrelevant to an assessment of probability.[60]

Naturally, Jeffreys did not understand Fisher's point. "With regard to the quotation from Keynes, I did not overlook Keynes's answer. I do not agree with it. If the word 'country' has any intelligible meaning, I think it is a very relevant piece of information whether Great Britain and Ireland are two countries or two parts of the same country." For Fisher, Jeffreys's formulation of the Principle of Insufficient Reason was arbitrary, and could lead to inconsistent results. But for Jeffreys it was Fisher who was inconsistent, for not recognizing that the relevant probabilities must be assessed relative to the same body of data. Keynes's apparent contradiction occurred because the same words used in different senses convey different quantities of background information. Yet a probability was always relative to a body of knowledge, and so would naturally be affected if the words used to express this knowledge were construed in different ways. Jeffreys remained mystified in his correspondence with Fisher. "My immediate reaction is that I cannot take it as obvious that people's linguistic habits are meaningless; but I think there is something else and can't see what it is."[61]

5.5.3. The definition of probability

Jeffreys had muddied the water with the Principle of Insufficient Reason. Well, Fisher had many trusty objections to inverse probability in reserve. First, the epistemic interpretation forms no basis for science. To be useful, a probability must admit of experimental verification, as can frequencies in gaming and genetic applications. This is not possible in Jeffreys's "subjective and psychological" version: "It will be noticed that the idea that a probability

---

[59] Jeffreys 1931, pp. 20, 29; Jeffreys 1932c, pp. 83–7; Jeffreys 1933a, pp. 528–9; Jeffreys to Fisher, 1 March 1934 (Bennett 1990, p. 153).

[60] Fisher 1934, p. 4.

[61] Jeffreys 1934, p. 14; Jeffreys to Fisher, 1 March 1934 (Bennett 1990, p. 154).

can have an objective value, independent of the state of our information, in the sense that the weight of an object, and the resistance of a conductor have objective values, is here completely abandoned."[62] Second, the epistemic interpretation relies on the unjustified equation of a logical relationship between propositions with the relative degree of belief in those propositions. Why should these have anything to do with each other? Keynes, though he had defined probabilities in terms of the addition and multiplication laws, was guilty of the same charge. Third, inverse probability is too blunt a tool for the many and various sorts of uncertainty that occur in practice. "However reasonable such a supposition may have appeared in the past, it is now too late, in view of what has already been done in the mathematics of inductive reasoning, to accept the assumption that a single quantity, whether 'probability' or some other word be used to name it, can provide a measure of 'degree of knowledge' in all cases in which uncertain inference is possible."[63]

Jeffreys was baffled. "At present it seems that the relations between us are of the form 'A thinks B has done all the things that B has been at special trouble to avoid'; where A = J, B = F and conversely."[64] Fisher had accused his definition of being unverifiable. Yet for phenomenalist Jeffreys it was Fisher's definition that, involving infinities, should be rejected as meaningless. "That a mathematician of Dr. Fisher's ability should commit himself to the statement that the ratio of two infinite numbers has an exact value can only be regarded as astonishing."[65] Jeffreys repeated the objections he had made with Wrinch in 1919. Talk of a limit, or any ratio of frequencies, begs the question. In any particular case, there is no guarantee that such a limit exists. Of course, the existence of a limit can often be inferred, as can its particular value. But since inference is a probabilistic process, the concept of probability must be more fundamental than frequency. "[T]he simple answer to Fisher's theory is that the hypothetical infinite population does not exist, that if it did its properties would have to be inferred from the finite facts of experience and not conversely, and that all statements with respect to ratios in it are meaningless."[66]

Jeffreys leveled the same objection at Fisher's claim that his frequency was an objective measure, like weight or density. Where could such frequencies – supposedly "independent of the state of our information" – come from, if not by inference from physical laws? Probability, in Jeffreys's inferential sense,

[62] Fisher 1934, pp. 3, 4.
[63] Fisher 1934, p. 8.
[64] Jeffreys to Fisher, 1 March 1934 (Bennett 1990, p. 152).
[65] Jeffreys 1933a, p. 533.
[66] Jeffreys 1933a, p. 533; see too Jeffreys to Fisher, 24 February 1934 (Bennett 1990, p. 150).

must *precede* any judgment about such frequencies, or 'chances' as Bartlett called them. Further, the frequency definition rested on the irrelevance of other factors. Yet this too must be assessed empirically, using the calculus of probability. For example, the probability of throwing a 6 with a die is $\frac{1}{6}$ only if the die is unbiased, uniformly free to rotate in any direction during flight, and the laws of dynamics hold. Each of these conditions is an approximation. Whether this approximation is good enough in any situation to consider the probability a fixed chance is something to be assessed using the theory of probability. Jeffreys: "I have a suspicion that a chance in [Bartlett's] sense exists only when inferred from physical laws, resting already on such evidence that other knowledge is practically irrelevant; in other words, that the existence of chances can be inferred only by using the general theory of probability."[67] He noted that in practice, Fisher obtained numerical estimates for his frequencies not by constructing infinite sets, but simply by counting cases each assumed to be equally probable.

Fisher, with no such commitment to probability as a primitive notion, found this difficult to understand. He was providing not a theory of learning but a statistical model for experimental situations. Such an abstraction, by definition, was not physically realizable. "I myself feel no difficulty about the ratio of two quantities, both of which increase without limit, tending to a finite value, and think personally that this limiting ratio may be properly spoken of as the ratio of two infinite values when their mode of tending to infinity has been properly defined," he wrote Jeffreys. Fisher had already, in his earlier papers, made the frequency definition rigorous and unambiguous.[68]

5.5.4. Logical versus epistemic probabilities

A similar standoff followed Fisher's criticism that Jeffreys had conflated the logical and epistemic versions of probability. As Fisher saw it, the logical relationship must measure something akin to the amount of 'information' a body of data had to offer relative to some proposition. This had nothing to do with the extent to which that data bore on the truth of the proposition, still less to a psychological degree of belief. Indeed, for Fisher, the failure of the Principle of Insufficient Reason to provide an objective grounding for inverse probability was doubly significant. Ignorance and the knowledge of

---

[67] Jeffreys 1934, pp. 15–16. Jeffreys doubted whether such objective probabilities existed: "Bartlett's notion of chance is an extreme case which is often approximately, but never, I think, exactly, realized in practice." Jeffreys 1934, p. 13.

[68] Fisher to Jeffreys, 26 February 1934 (Bennett 1990, p. 151). For a rigorous frequency definition see, e.g., Fisher 1925a.

ignorance are different things. Not only does this distinction signal that any attempt to use the Principle of Insufficient Reason is futile and misplaced, it also indicates "that an initial mistake is introduced in all such definitions as that of Jeffreys in assuming that the degree of knowledge or rational degree of belief is, in all cases, measurable by a quantity of the same kind."[69] Fisher's probability was a unique measure, fixed by experimental circumstances. Thus he saw a distinction between the form of uncertainty represented by the probability itself, and that arising from our imperfect knowledge of its value. The two quantities he saw confused in Jeffreys's interpretation – weight and truth-content of information – would have clearly different effects on these uncertainties. New information would always reduce the uncertainty of our knowledge of a probability, but could lead to a reassessment of the value of the probability in either direction. By definition, uncertainty is sometimes a vague matter. Yet in Jeffreys's view all probabilities were precise.

Since Fisher's probabilities obeyed the laws appropriate to games of chance, there was no reason to suppose them sufficient to express every form of uncertainty. "Even if, in a logical situation providing a basis for uncertain inference, we confine attention to quantitative characters, measuring the extent to which some inferences are to be preferred to others, different situations, among the kinds which have already been explored, provide measures of entirely different kinds. Thus, a knowledge of the construction and working of apparatus, such as dice or roulettes, made for gaming, gives a knowledge of the probabilities of the different events or sequence of events on which the result of the game may depend. This is the form of uncertain inference for which the theory of probabilities was developed and to which alone the laws of probability are known to apply." Fisher provided separate measures for the ideas he saw confused in Jeffreys's definition, and again pointed out that prior probabilities are not needed to produce measures of inference. Maximum likelihood is ideal for problems of estimation, and "[i]t is to be anticipated that a detailed study of logical situations of other kinds might reveal other quantitative characteristics equally appropriate to their own particular cases."[70]

For Jeffreys, with his understanding of probability as primitive, this objection to probability as an insufficient gauge of uncertainty made no sense. A probability measured belief in some proposition relative to a specified body of knowledge. New information, therefore, did not reduce the uncertainty in an unknown probability, but required the calculation of a wholly new probability. Jeffreys regarded individual knowledge as more fundamental than

[69] Fisher 1934, pp. 5–6.
[70] Fisher 1934, pp. 6, 8.

statements about the world, and took as axiomatic that such knowledge could be quantified as a degree of belief by a probability. That was the whole point of the concept. Fisher's assumption of a limiting ratio, like his discussion of types of uncertainty, put everything the wrong way around. Probability was just the quantity used to *measure* knowledge such as that pertaining to dice or roulettes. Jeffreys's probability distributions were *not* different sorts of thing from the relevant quantities arising with dice – or indeed, genetic experiments – since those probabilities were themselves the outcome of observation. "My attitude is that you are asking the wrong question . . . I think you are regarding a probability as a statement about the composition of the world as a whole, which it is not and on a scientific procedure could not be until there was nothing more to do."[71] Jeffreys groped for elucidation. Many opponents of his theory were unwilling to accept that two propositions could be related with a definite value of probability. Perhaps Fisher's introduction of the idea of information was related? Yet a probability relating two propositions could always be defined: this followed from his and Wrinch's postulate that probabilities could be arranged in numerical order.[72] And on this definition a probability is not the same thing as relevance. "Thus we may have $P(p|q) = P(p|qr)$ for a wide range of values of r; then r, within limits, is said to be irrelevant to the probability of p given q."[73]

5.5.5. Role of science: inference and estimation

Jeffreys and Fisher could not reach agreement on these issues because in each case they were based on separate ideas of the proper role of science. Two examples will make this clear. The first concerns the matter of objectivity. To Fisher's criticism that epistemic probability was not objective, Jeffreys had replied that this word, if it meant anything at all, referred to a condition that could only be inferred in consequence of our investigations. Fisher's accusation was largely a matter of rhetoric. For force, it relied on his belief that a "subjective and psychological" interpretation of probability was inherently unscientific. Yet for Jeffreys, science is just a matter of individual experience. Of course it is subjective! "The distinction between subjective and objective has been a matter for argument between philosophers for some 2000 years, without, I think, appreciable approach towards agreement. For

---

[71] Jeffreys to Fisher, 10 April 1934 (Bennett 1990, p. 160).

[72] That a numerical probability relating two propositions could always be assigned, Jeffreys pointed out, also followed from the separate approaches, based on expectation, of Bayes and Ramsey. Jeffreys 1934, p. 12.

[73] Jeffreys 1934, pp. 12–14; see also Jeffreys to Fisher, 1 March 1934 (Bennett 1990, p. 153).

scientific purposes, our experience reduces on ultimate analysis to sensations, which are largely subjective on almost any philosophical system. It is a fact that different people interpret their essentially private sensations in terms of the same 'reality,' but I think I have shown in my book that this interpretation involves the whole procedure of inference, which is therefore more fundamental in knowledge that any question of reality, which is not very important for scientific purposes."[74]

The second example concerns Jeffreys's rejection of the frequency definition. Not only was such a definition unavailable, but for Jeffreys, Fisher's frequencies were not even the relevant scientific quantity. Scientists are not interested in the properties of some hypothetical population, he argued. They are concerned about particular cases. Yet Fisher said nothing about how to transfer a probability statement from a population to an individual; indeed, he said this process was illegitimate. As Jeffreys wrote Fisher, "suppose that by your methods you compare two methods of growing potatoes and show that in 90 per cent of cases method A will give a greater yield than method B. Suppose a farmer asks you 'What reason is there to suppose that *I* will get a better yield by method A?' It seems to me that your only answer is in terms of the view that probability is intelligible without definition."[75] That is, probability – in Jeffreys's epistemic version – is useful just because it is that concept which applies to belief in particular cases or the next occasion rather than collectively to an ensemble.

This relates to the logic of statistical inference. Fisher's account centered on the significance test, and was based on a form of counterfactual reasoning. We reject the null hypothesis because if true, the results we obtained, or those more extreme, would be highly improbable compared to other possible outcomes of the experiment. But for Jeffreys an inference is only relevant if it can be applied to particular cases. It is not a matter of testing reliability over repeated samples. He objected to Fisher's significance test because it depended on the tail probability of the test statistic, a value almost entirely determined by outcomes other than that actually observed. (As he was memorably to put it, *"a hypothesis that may be true may be rejected because it has not predicted observable results that have not occurred."*[76]) Jeffreys had a similar objection to fiducial probability. The concept was only intelligible to him when interpreted as the distribution of the *probability* of the standard deviation in the sample, rather than the distribution of the standard deviation in the infinite

[74] Jeffreys 1934, p. 13.
[75] Jeffreys to Fisher, 1 March 1934 (Bennett 1990, p. 153).
[76] Jeffreys [1948], p. 357.

population. Even so, it "refers to a mean taken over all possible random samples, and it may be asked why this should be thought to have much relevance to any particular sample."[77] Fisher had pointed out that whereas the fiducial distribution refers to the "population of all random samples, that of the inverse probability is a group of samples selected to resemble that actually observed."[78] But as far as Jeffreys was concerned, we were only interested in the data that we actually had. Why should we care about a distribution averaged over all possible samples? "Fisher's integration with respect to all values of the observed values thus involves a fundamental confusion, which pervades the whole of his statistical work, and deprives it of all meaning. The essential distinction in the problem of inference is the distinction between what we know and what we are trying to find out: between the data and the proposition whose probability on the data we are trying to assess." We have our observations. "To integrate with respect to them and average a function of them over the range of integration is an absolutely meaningless process. Yet in Fisher's constructive, as well as in his destructive work, this process is carried out again and again."[79] Nor is the supposed independence of the fiducial probability on a prior distribution an advantage. It makes inferences from the sample impossible. If, as is often the case, we *do* have some relevant background knowledge – Jeffreys gave a balls-in-urn illustration – the fiducial distribution is worse than useless for the purpose of inference. "The theory of probability, on the other hand, allows us to take such previous knowledge into account completely."[80]

The correct form of probability depended finally on the sort of science practiced. For Jeffreys, as we have seen, this was primarily a process of weighing and fitting sometimes sparse data from various sources in order to perform extrapolations and gauge their reliability. Did the data indicate a new physical mechanism at work, or was the scatter merely experimental noise? "The sort of thing that bothers me is this," he wrote Fisher. "In seismology we get times of transmission to various distances, and fit a polynomial of degree 3, say, to them. The significance of the last term really involves the probability that such a term will be present. The usual thing is to keep it if it is some arbitrary multiple of its standard error, but I think it ought to be possible to frame a rule with *some* sort of argument behind it . . . " He gave another instance of the role of probability theory in the same letter. "I want some time to tackle

[77] Jeffreys 1933a, p. 534.
[78] Fisher 1933, p. 348.
[79] Jeffreys 1933a, p. 532.
[80] Jeffreys 1933a p. 534. See too p. 533; Lane 1980, p. 158.

the problem of the distribution of gravity over the Earth. The trouble is that the stations are very irregularly distributed and there is no question of being able to interpolate directly and work out coefficients of spherical harmonics by integration. The nearest I can see to a practical method is to assume a series of harmonics up to order 4 and try to fit it to observations. But then the harmonics of order 5 may have the same sign over most of the United States, and it would be ridiculous to treat each station there with the same weight as a single station in the Antarctic Continent. Some kind of grouping seems indicated; the question is what would be the best area to include in a group?"[81]

Since he took probability as the fundamental concept that described inference, Jeffreys was concerned to emphasize the necessity of *a priori* elements for the theory to get further than mere description of the data. He had said the same in his commentary on the Haldane–Fisher exchange: the point of sampling is to move from the sample to a statement of the class, and this *cannot* be done without background assumptions. Indeed, for efficiency it *should not* be done without such assumptions: if, for example, we already suspect a general law of one sort or another, we will be able to encode this in our prior probability and hence draw more confident inferences from fewer observations. This was the other side of the argument that the influence of prior probabilities declined with increasing quantities of data.[82] Many inverse probabilists made this point only in defense, to excuse their use of priors. This left them vulnerable to the criticism Fisher had made in his response to Haldane: if the prior probability is truly unimportant for large samples, it is irrelevant and should have no part in our reasoning; if it affects our reasoning, then it is misleading.[83]

---

[81] Jeffreys to Fisher, 21 March 1934 (Bennett 1990, p. 156); the second quote, not included in Bennett, is from the copy of this letter in HJ.

[82] In *Scientific Inference*, Jeffreys had noted that in practice we swamp the uncertainty in the prior probabilities by repeated verification and crucial tests: "The scientific man might, if he took enough trouble, evaluate the prior probability accurately; but in practice he is not interested in the accurate evaluation of a moderate probability. He prefers to obtain such additional information as will make the posterior probability approach impossibility or certainty whatever the prior probability may have been; and when that is done he no longer needs to evaluate the prior probability." Jeffreys 1931, p. 23.

[83] If the prior probability "is in reality irrelevant to our conclusions, it should have no place in our reasoning; and . . . if the form of our reasoning requires its introduction, the fault lies with our adoption of this form of reasoning. The conclusion to be drawn from the decreasing importance of our *a priori* information is not the trivial one that, by introducing false *a priori* data, we may quite possibly not be led far astray, but, rather, that it indicates the fundamentally different position that conclusions can be drawn from the data alone, and that, if the questions we ask seem to require knowledge prior to these, it is because, through thinking only in terms of mathematical probability, and of the deductive processes appropriate to it, we have been asking somewhat the wrong questions." Fisher 1932, pp. 258–9.

But Jeffreys was not apologizing for the use of priors, instead pointing out that they were necessary for inference. True, prior probabilities were often awkward to quantify and were usually unimportant for large samples. But this did not warrant their neglect. Statisticians, he suggested, took large samples to reduce the probable error of their inferred values, not because the use of prior probabilities was illegitimate. Nor should it be a concern that such *a priori* elements cannot be proved. As Jeffreys repeated, he was not in the business of trying to *prove* induction logically or experimentally. Such a proof was an impossibility. Instead, he was looking for consistency: the theory must "agree with ordinary beliefs about its validity."[84] The Simplicity Postulate and his form of *a priori* probability for qualitative laws were attempts to account for the inference of general laws from the data of real scientific practice, when the sampling model is no good and equally possible alternatives not immediately available. "By their very nature these rules cannot be established by experience: they must be judged by their plausibility, the internal consistency of the theory based on them, and the agreement or otherwise of the results with general belief."[85] "This point needs emphasis," he wrote in his second paper, "because many critics of the theory of probability think that they can dispose of it by pointing out that it makes assumptions that cannot be proved either logically or experimentally. Any theory capable of justifying generalization must necessarily make some assumptions of this character, and the issue is whether such assumptions shall be made or generalization entirely abandoned; and, of course, science with it."[86]

Jeffreys expanded the argument in his letters to Fisher. The Method of Maximum Likelihood is useful because it is proportional to the posterior probability given by Laplace's theory. But it cannot serve as a method of inference. On its own, the Method of Maximum Likelihood would suggest an exact form for an inferred law. And the data alone can always be exactly fit by an infinite number of laws. The likelihood therefore *cannot* be all that is needed for inference. Neither can pure logic help. The likelihood is a measure of the sample alone; to make an inference concerning the whole class we combine the likelihood with an assessment of prior belief using Bayes's theorem.[87] Inverse probability has the virtue that this necessary prior knowledge is expressed overtly, rather than hidden in the specification. Surely Fisher must acknowledge that maximum likelihood is not enough to account

---

[84] Jeffreys 1934, p. 9.
[85] Jeffreys 1934, p. 9.
[86] Jeffreys 1933a, p. 525.
[87] Jeffreys 1933a, pp. 527–8.

for scientific practice? "It would be a quite logical position to maintain that all quantitative laws are merely interpolations and that all values of y for other than the observed values of x are equally likely; e.g., that the predictions in the *Nautical Almanac* are meaningless; but if you really mean that, I think you ought to say so."[88] Again, *a priori* postulates are not to be despised. "I cannot help it, but there is a general belief in the possibility of establishing quantitative laws by experience, and I am not prepared to say that the general belief is wrong. I think I have stated a postulate that expresses it sufficiently clearly for practical purposes, but the postulate cannot be proved experimentally. But since such a postulate must from its nature be believed independently of experience that is a recommendation. What I want is, since an *a priori* postulate is needed anyhow, to choose it in such a way that the maximum number of alternatives are left for experience to select from."[89]

Jeffreys's emphasis on *a priori* postulates for inference seemed to Fisher to miss the mark. He was concerned not so much with inference than the direct analysis of abundant or at least well-behaved data. The rules and methods he had developed for estimation were prescriptive rather than descriptive, and appropriate to specific experimental circumstances. (Recall the quote a few pages back: "Thus, a knowledge of the construction and working of apparatus, such as dice or roulettes, made for gaming, gives a knowledge of the probabilities of the different events or sequences of events on which the result of the game may depend. This is the form of uncertain inference for which the theory of probabilities was developed, and to which alone the laws of probability are known to apply.") Fisher outlined the correct procedure. In the theory of estimation "we are provided with a definite hypothesis, involving one or more unknown parameters, the values of which we wish to estimate from the data. We are either devoid of knowledge of the probabilities *a priori* of different values of these parameters, or we are unwilling to introduce such vague knowledge as we possess into the basis of a rigorous mathematical argument."[90] Fisher said more on what he was aiming at in his correspondence with Jeffreys: "You say I object to the introduction of an *a priori* element, [so] I should like to get this notion a little more precise. What I object to is the assertion that we must, in considering the possible values of an unknown parameter used to specify the population sampled, introduce a frequency distribution or a probability distribution of this parameter, supposedly known *a priori*. If we really have knowledge of this kind, as in some problems,

[88] Jeffreys to Fisher, 24 February 1934 (Bennett 1990, p. 150).
[89] Jeffreys to Fisher, 1 March 1934 (Bennett 1990, p. 154).
[90] Fisher 1934, p. 7.

which can be reconstructed with dice or urns, I do not deny that it should be introduced, but I say that we often do not possess this knowledge, i.e., either that we are in absolute ignorance or that our knowledge is so vague or unsatisfactory that we may properly prefer to keep it in reserve and see what the observations can prove without it and whether it, for what it is worth, is confirmed or contradicted by the observations. I claim, in fact, that it is at least a legitimate question to ask: 'What can the observations tell us when we know the form of our population, but know nothing of the *a priori* probability of the different values that its parameters may have?"[91]

Yet Fisher conceded that some *a priori* knowledge is necessary. The above quote continues: "This is the situation which I treat as the typical one in the Theory of Estimation, but it would be quite legitimate to say that our assumed knowledge of the form of the distribution, which is needed before there can be anything to estimate, is in the nature of *a priori* knowledge. In fact, such knowledge seems to me essential in what I call problems of specification, but out of place in the next stage, when problems of estimation arise." Thus while maximum likelihood was the proper way to fit data to some specified curve, Fisher agreed that experience could be used with advantage to revise the specification or form of curve. Statistical science, however, is an objective matter, properly restricted to the facts. The problems of specification and estimation are logically distinct, and while Fisher admitted pragmatic considerations to the former, he insisted that the main body of quantitative scientific work could and should be kept separate.[92] After all, the experience guiding the choice of specification is often of a vague kind. Presenting such information in numerically exact form is dishonest. Even were it easily quantifiable, we would have no wish to mix such indefinite quantities with the direct data of experiment. Specification, then, is not governed by rules nor subject to a rigorous mathematical calculus. Jeffreys had tried to describe this process in terms of probability, but it is not clear "that probabilities provide our only, or chief, guide in this matter. Simpler specifications are preferred to more complicated ones, not, I think, necessarily because they are more probable or more likely, but because they are simpler. As more abundant data are accumulated certain simplifications are found to be very unlikely, or to be significantly contradicted by the facts, and are, in consequence, rejected; but among the theoretical possibilities which are not in conflict with any existing body of fact, the calculation of probabilities, even if it were possible, would

[91] Fisher to Jeffreys, 29 March 1934 (Bennett 1990, p. 158).
[92] Fisher 1934, p. 7.

not, in the writer's opinion, afford any satisfactory ground for choice."[93] In private, Fisher was more accommodating. Yet he made clear that the pragmatic considerations of specification lay outside the domain of objective science. "In practice, as we know, simplicity is sought in specification, but whether it is to be justified primarily by convenience, subject to the satisfaction of tests of goodness of fit or whether, as you have proposed, it should more properly be justified in terms of probability, I have not any strong opinion."[94]

Fisher and Jeffreys ultimately disagreed over the domain of statistical science. One consequence of the high status of theory, augmented perhaps by more visceral tendencies to generalize, is the tendency common in system-builders and methodists to regard the lessons of their specific area of study as more widely applicable. Thus Jeffreys took his treatment of induction in geophysics as a general picture of learning from experience; and Fisher his model of genetics and agronomic design as a standardized form of doing science. Bartlett's gloss reflected a third interpretation, one natural for a statistician rather than primarily a practicing scientist, concerned to demarcate an area for his new discipline. As he saw it, statistics like any science starts by refining and making precise terms used in everyday language. Jeffreys's observation that generally the word 'probability' is used to refer both to events and hypotheses, therefore, was neither here nor there. The study of statistics is properly and most profitably confined to chances – that is, to situations in which the model of objective frequencies can be applied as a good approximation to countable events occurring in the world. Only then are precise statements possible. (*Pace* Jeffreys, we need not necessarily take such statistical qualities as anything more than theoretical constructs; good fits to data should not be hastily interpreted in terms of ultimate causes.) Bartlett consequently saw statistical inference as a subset, not the whole, of scientific inference, and this in turn as part of the broader problem of general inference. Jeffreys's attempt to explain all inference in terms of probability, whilst like all general theories superficially seductive, was an over-simplification that introduced more difficulties that it explained. Inverse probability was a wretched hybrid of exact chances and unknown priors for which a formal treatment is not possible. And Jeffreys's solution to the problem of induction through the deduction of objective priors was based on convention; where they appeared convincing, his methods relied on the support of objective statistics as prior probabilities. "Inverse probability, in its widest sense, is implicitly contained in an inductive inference in any science, but that is no reason why, because the particular

---

[93] Fisher 1934, p. 8.
[94] Fisher to Jeffreys, 29 March 1934 (Bennett 1990, p. 158).

probabilities known as chances form a large part of the subject-matter of statistics, it should be used explicitly, and entangle these chances with prior probabilities and previous or *a priori* knowledge."[95]

Fisher and Bartlett thus agreed on the proper role of the statistician.[96] He was not anxious to leap to conclusions nor satisfied with mere rules for consistency. Instead, he was concerned to analyze the quality and relevance of the data and the assumptions of his statistical model. Was the sample representative? Were there any indications that the experimental conditions had altered? This was not a matter for inverse probability. Instead, the method of maximum likelihood, using direct calculations of chances, should be used to generate optimum values of parameters. Only after he had addressed the data did the statistician bring his professional skills to bear on the matter of inference. This required judgment and common sense. Jeffreys reduced everything – data, experience, psychological factors – to a single number and thus addressed the wrong sorts of question in too final a form. The professional statistician, in contrast, carried a number of sophisticated tools for uncertain inference, each appropriate in different cases. A scientist with previous suspicions about a parameter wanted not some measure of his belief in a particular value, but to know whether the sample fit with the assumptions of his model. In this case, a significance test would ascertain the chance of the apparent discrepancy of the optimum value. Objective probable limits could then be provided by Fisher's fiducial method.[97]

5.6. CONCLUSIONS

The Fisher–Jeffreys exchange was, as Jeffreys later wrote, marked by confusion on both sides. Despite similarities in terminology, Fisher and Jeffreys were asking different questions and assuming different modes of inquiry. The exchange of letters did little to clarify the matter. Jeffreys repeatedly commented that he did not fully understand the objections to his theory. ("I find it difficult to answer criticisms of the theory, because most of them seem to

---

[95] Bartlett 1933, p. 534; MSB. Bartlett was deeply impressed by Fisher's reformation of statistics as a practical rather than purely academic study, and moved in October 1933 to work in the new statistics department at University College London, arriving shortly after Fisher took up his chair in eugenics.

[96] Fisher's conversion of Jeffreys's precision constant to a standard deviation "for the convenience of the majority of statisticians" was probably a dig at Jeffreys's amateurishness as a statistician. Fisher 1933, p. 343. Jeffreys's use of h followed Eddington's treatment, presented in the lectures on the combination of observations that Jeffreys had attended as an undergraduate in 1913.

[97] Bartlett 1933, pp. 533, 534.

refer to something different from what I intended, and I cannot see what."[98])
For his part, Fisher regarded Jeffreys's explanation of his approach as more
reasonable than in the original 1932 paper, but although muting the tone of
his criticisms, left their substance chiefly unaltered in his rejoinder.[99]

Jeffreys thought Fisher was claiming to supply a rival probabilistic theory of knowledge. This led Jeffreys to open his second paper, like *Scientific
Inference*, with a discussion of approaches to inference. He described Russell
as typical of contemporary philosophers in claiming to believe that induction is either "disguised deduction or a mere method of making plausible
guesses." Such sterile argument, however, was a luxury that practicing scientists could ill afford. Even so, there was little agreement on inference among
scientists. Eddington, at one extreme, held that all science is *a priori*. Like
other 'official' relativists, he claimed to believe Einstein's theory on purely
philosophical grounds, rather than because of the experimental verifications
and the fact that – as Jeffreys had troubled to show – alternative explanations,
involving matter near the sun or other laws of gravitation, could not account
for the observations. "To holders of such a view experiments become merely a
game that some people . . . find amusing, and sometimes a means of enabling
them to detect mistakes in the mathematics."[100] Jeffreys saw Fisher as standing at the opposite extreme, and claiming to avoid *a priori* considerations
altogether. He wrote to Fisher, "I prefer to have just enough *a priori* material to give experiment a chance to decide what is true. You give too little;
Eddington gives too much."[101]

Jeffreys was repeatedly critical of the frequency theory as smuggling in
concealed and unsupported assumptions, and thus falling down according to
his then-fashionable view that such a theory of scientific inference, though by
nature unprovable, should be reduced to as few axioms as possible.[102] Nav-
igating correctly between the Scylla of abandoning attempts to infer beyond
the data, and the Charybdis of assuming that experience is unnecessary for
knowledge, required a recognition that science was inherently an individual
and inferential enterprise, to be described with an immediate degree-of-belief
probability. It could be grounded neither wholly *a priori* nor on frequencies.
"I have no *a priori* certainty that the number of electrons and protons in the

[98] Jeffreys 1934, p. 10. See also Jeffreys 1934, pp. 12, 13–14; Jeffreys to Fisher, 24 February
1934 (Bennett 1990, p. 150); Jeffreys to Fisher, 1 March 1934 (Bennett 1990, p. 153).
[99] Fisher to Jeffreys, 29 March 1934 (Bennett 1990, p. 156).
[100] Jeffreys 1933a, p. 526. See also Jeffreys 1934, p. 12; Jeffreys to Fisher, 24 February 1934
(Bennett 1990, p. 150); Jeffreys to Fisher, 1 March 1934 (Bennett 1990, p. 154).
[101] Jeffreys to Fisher, 24 February 1934 (Bennett 1990, p. 150).
[102] Jeffreys 1933a, pp. 532–3; Jeffreys to Fisher, 1 March 1934 (Bennett 1990, p. 152).

universe is equal, or that the principle of general relativity is true; my respect for both beliefs is due to the fact that they agree better with experience than various other alternatives that I might consider equally likely *a priori*. In the former case I do not consider an indefinite repetition of universes with all ratios of the numbers of protons and electrons equally well represented; this would be as much an *a priori* idea as probability itself, and therefore offers no advantage. I simply state my previous ignorance of the composition, and proceed to consider the consequences of observational data in modifying this ignorance."[103]

Yet Fisher was constructing not a formal description of physicists' behavior, but a prescriptive theory that governed the acquisition and mathematical treatment of statistical data. This was a normative view, in which science was a matter of random statistical aggregates, and the data representative of a population. The statistician was to analyze this data, and use it to answer well-posed questions. Estimation and significance-testing made up the proper scientific procedure. Naturally, the scientist was also interested in moving from the particular to the general, but such inference, when appropriate, consisted in reasoning from a sample to a pre-specified class rather than from a set of observations to a general law. Fisher saw his work and the associated frequency interpretation in the tradition of statistical science. Had not Bayes himself based probabilities on expectations, and hence frequencies? This moreover was a natural interpretation historically, since "expectations, as of cargoes at sea or contingent rights in property, must have been bought and sold long before any value placed upon them was formally analysed into the two components, probability of possession and value if possessed."[104] Hence Fisher was not prepared to countenance Bartlett's suggestion that the term 'chance' be reserved for statistical probabilities. "Since the term 'mathematical probability' and its equivalents in several foreign languages have been used in this sense and almost exclusively in this sense, for over 200 years, it is impossible to accept Mr. Bartlett's suggestion, in his thoughtful discussion of the topic, that only the word 'chance' should be used for the objective probabilities with this meaning, and that the word 'probability' should be confined to the recent and perhaps ephemeral meaning which Dr. Jeffreys has assigned to it."[105]

Each of Jeffreys and Fisher was too wedded to his interpretation to appreciate the aims and assumptions of the other man. Thus Fisher in his final

---

[103] Jeffreys 1934, p. 12.

[104] Fisher to Jeffreys, 13 April 1934 (Bennett 1990, p. 160). As discussed in Chapter 3, Fisher idiosyncratically read Bayes's interpretation of probability as based on frequencies in a hypothetical infinite population. See Fisher 1936, p. 451.

[105] Fisher 1934, p. 7.

paper continued, albeit rather weakly, to use the language of frequency to insist on ambiguity in Jeffreys's original formulation of the one-third argument. And to Jeffreys's complaint in his second paper that "Fisher proceeds to reduce my theory to absurdity by integrating with respect to all values of the observed measures," Fisher acidly replied that this he was not inclined to deny.[106] For his part, Jeffreys was unwavering in his refusal to consider a restricted theory of statistical probability. Indeed, thinking that Fisher's theory was a disguised form of inverse probability appropriate for large samples, he was more concerned by a subtlety of Bartlett's that seemed to invalidate the one-third argument. Bartlett had pointed out that Jeffreys's original formulation – since the middle of the three values is equally likely to be the first, second, or third made, the probability that the third measurement lies between the first two can be set equal one-third – involves two distinct statements. The probability that the middle in magnitude of three observations was the third made is indeed one-third. "But this tells us nothing about the probability that, given two observations, a *third* observation will be the *middle* one; as we have seen, if we know the range of error, this probability will be a function of the values of the first two observations, if we do not know the range of error, its value is still undetermined."[107] Bartlett's comment highlighted an inconsistency in Jeffreys's defense of dh/h: that this form of prior probability remained unchanged – as it should – after a single reading, but was modified after a second to reflect that the difference in readings suggested a scale for the precision (and the mean a clue to the true value). With two measurements in hand, was not this the situation at the start? After a worrying period, Jeffreys applied the Principle of Insufficient Reason to convince himself that the one-third argument still held in the case of two particular measurements. The difference between the first two observations still gives no information about whether the third should lie between; this is equivalent to saying that the scale chosen for measurement is irrelevant.[108]

The differences in temperament – Jeffreys placid but implacable, Fisher volcanic and paranoid – would seem to have called for a bruising encounter.

---

[106] Jeffreys 1933a, p. 532; Fisher 1934, p. 3.

[107] Bartlett 1933, p. 527.

[108] Jeffreys 1934, p. 14; Jeffreys to Fisher, 1 March 1934 (Bennett 1990, p. 155). Jeffreys's letter of 20 April, 1934 continues: "I got worried the other day because of the probability of an [observation] given the time value is *anything* of [a general form], my arguments about the prior [probability] of h seemed to hold just as for the normal law, and it did not seem obvious that dh/h could still imply the 1/3 business. But after a rather nasty triple integration I find that it is perfectly general. I don't want to publish it at once as I may develop more definite ideas about some other things and they might as well go together."

Yet despite their inflexibility, the tone of the correspondence was cordial, almost friendly. Jeffreys's struggle to grasp Fisher's objections, like his appeals for elucidation, was genuine rather than rhetorical. Fisher in return was conciliatory, not abusive. Jeffreys and Fisher had mutual respect as fellow scientists. Unlike their usual opponents – armchair philosophers and narrow statisticians of the mathematical school, respectively – each recognized in the other a prime concern with the development of probability theory for practical use, rather than for philosophizing or mathematical pedantry for its own sake. Their letters discuss many specific problems they were working on.[109] Jeffreys, in particular, was anxious to make clear that his quarrel was less with Fisher than with those, like Eddington, who denied the importance of probability altogether. In purely numerical terms, Jeffreys pointed out, he and Fisher would only differ in exceptional cases.[110]

Jeffreys took inverse probability for granted, and took Fisher's method of maximum likelihood to be a version with the assumption of uniform prior probabilities. This, he conceded, was usually a valid assumption for statisticians like Fisher, chiefly concerned with large samples. ("So far as I can see, this covers all the cases that Fisher and other statisticians consider, and I should be the last person in the world to deny the utility of approximation."[111]) Hence Jeffreys saw no reason not to study likelihood. It was "usually a fair assumption to the truth when statistical methods are actually used, and it would not be prohibitively difficult to estimate the correction for the effect of

---

[109] Many of these discussions are not included in the selection printed in Bennett 1990.

[110] Jeffreys to Fisher, 24 February 1934 (Bennett 1990, p. 150). Jeffreys wrote in the *Theory of Probability* [1948], pp. 364–5, that the differences between them had been "exaggerated owing to a rather unfortunate discussion some years ago, which was full of misunderstandings on both sides . . . . This discussion no longer, in my opinion, needs any attention. . . . I have in fact been struck repeatedly in my own work, after being led on general principles to a solution of a problem, to find that Fisher had already grasped the essentials by some brilliant piece of common sense, and that his results would either be identical to mine or would differ only in cases where we should both be very doubtful." See also Jeffreys to Fisher, 18 May 1937 (Bennett, p. 162), and Jeffreys 1937, p. 1004, in which he writes that both theories are "satisfactory in the senses that (1) they do not treat any hypothesis as *a priori* certain; (2) they provide methods of choosing between hypotheses by means of observations; (3) they give estimates of the parameters involved in any hypothesis consistent with the observations; (4) they recognize that decisions made by them will sometimes be wrong, and give estimates of, alternatively, what degree of confidence we can reasonably attach to a decision, or how often, supposing that we use the rules regularly, we may expect the decision to be right; (5) they all contain provision for correcting wrong decisions when observations become capable of giving further tests, thus recognizing that scientific progress is by successive approximation."

[111] Jeffreys 1934, p. 10.

a different distribution of prior probability. It is only for small samples that the correction will be important."[112] He insisted nevertheless that in some geophysical cases, the data was so scanty that the priors would not always wash out, but have an important influence. ("Your expressions in ordinary cases are good approximations to the posterior probability; if I claim in one case to have got a slightly better one I don't see that you need worry."[113])

Fisher too was irenic. He suggested at one point that he and Jeffreys each tackle the same specific problem and publish the results to indicate the difference in methods. He proposed an example from genetics: an experimentalist assumes that a number of sub-types of a species of plant occur with frequencies in the ratio of powers of some unknown constant. He takes a random sample of ten plants from a wild population and finds three of one type, two each of a further two types, and one each of another three types. Can he say anything about the probable value of the constant? Jeffreys would doubtless, Fisher speculated, introduce a function representing the prior probability of the constant. But Fisher thought that in this case the likelihood for any value of the constant, calculable from the direct chance of obtaining the given partition of observations at that value, could serve as an objective guide for inference about the actual value of the constant. This could be used to calculate the probability that the next plant sampled will be of a wholly new kind.

Although this problem involved both probability and likelihood, Fisher used both terms to refer to unknown but fixed chances in a population of well-behaved and precisely specified data. Perhaps predictably, Jeffreys misunderstood. Was this a mutation problem? If so, would one not simply assume that the prior probability of the probability of mutation was uniformly distributed? If not, where did the scientist get the assumption of the ratios from? "I think if he knows this he must know a lot of other things too . . . "[114] Jeffreys proposed a different question, one that characteristically involved an insecure inference based on much vaguer information. A man arrives at a railway junction in a foreign town. He sees a tramcar numbered 100. Knowing nothing about the size of the town, but assuming the tramcars in the town are numbered consecutively from 1, that all are in service, and that each is equally likely to be observed, can he infer anything about the total number of tramcars? Jeffreys thought the problem was significant, and had "a feeling that there

---

[112] See also Jeffreys 1934, p. 10; Jeffreys to Fisher, 24 February 1934 (Bennett 1990, p. 150).

[113] Jeffreys to Fisher, 24 February 1934 (Bennett 1990, p. 150).

[114] See Fisher to Jeffreys, 26 February 1934; Jeffreys to Fisher, 21 March 1934; Fisher to Jeffreys, 29 March 1934; Jeffreys to Fisher, 10 April 1934; Fisher to Jeffreys, 13 April 1934 (Bennett 1990, pp. 151–2; 155–6; 157–8; 159; 160).

were probably about 200."[115] The problem could be tackled by assuming a distribution of prior probability for the number of tramcars.

Jeffreys posed the question to Bartlett, who was irked by the trifling nature of this problem. Had Jeffreys not read Fisher's 1930 paper on inverse probability or his work on fiducial intervals? The question was meaningless.[116] It might be expected that Fisher too would be irritated at such a question. He replied, however, that when reworked into continuous form, the tram problem admitted of a clear fiducial argument. Consider a parachutist attempting to infer the radius of a (circular) city, having landed near a stone marked '1 km. from city center.' "He then has an unambiguous fiducial argument as follows. 'If the radius of the city exceeds R kms., the probability of falling on or within the 1 km. circle is $< 1/R^2$. If this event has happened the fiducial probability that the radius exceeds R kilometres is therefore $1/R^2$ and the fiducial probability of it lying in the range dR is $2dR/R^3$. The fiducial median city has a radius $\sqrt{2}$ kilometres and an area of 2π. The fiducial mean radius is 2 km. and the fiducial mean area is infinite. The most likely R for the city however, is 1km., for this maximises the probability of his observation.'"[117] This did not work for the tramcars, because differentiation is not possible in the discrete case. (Fisher suggested that it was the concentration of earlier writers on discontinuous observations that had hidden the fiducial argument.) Yet Fisher's reworking of the problem again reveals a difference in aims. Jeffreys point was not to 'prove' or justify his 200 result, but to use it as a guide in the formulation of rules for consistent behavior. It was not that 200 is a reasonable inference for the number of tramcars, but merely that if people *did* have a feeling about the number 200, as his inquiries indicated – even Bartlett admitted so – a consistent distribution of their unconscious assignment of prior probability could be determined. This decreased as the reciprocal of the number of tramcars.[118]

These examples demonstrate once more the importance of disciplinary practices on the objectives of each man. Even though prominent for his work in small samples, Fisher came from a statistical tradition that found an easy association of probability with the frequencies fixed in huge quantities of data. Thus had Galton approximated his large family trees directly as infinite populations. Practical work, such as agricultural experiments, might not yield data that even in principle was infinitely repeatable. Yet the notion of proba-

[115] Jeffreys to Fisher, 10 April 1934 (Bennett 1990, p. 159).
[116] MSB.
[117] Fisher to Jeffreys, 13 April 1934 (Bennett 1990, pp. 160–61).
[118] In the Neyman–Pearson confidence interval approach, 200 comes out as the 50% risk solution.

bility as a frequency in a population made clear that in such an experiment the acquisition of data was to be considered at least an act of random sampling from a 'hypothetical' infinite population. This population was exactly specified, and produced self-contained sets of data that obeyed some law of chance. The data was in all cases comparable; thus experiments could be designed expressly to address well-posed problems.

Jeffreys's relationship with his data was wholly different. He worked with observations rather than experimental results, and with little control over these observations, had to make do with what he received from field workers or read in journals.[119] Such geophysical data was sparse and often of poor quality, requiring the extra information of prior probabilities in order to squeeze out meaningful conclusions. Jeffreys also needed to combine observations from a number of different sources. His inferences often tied geological and archeological results to astronomical observations and even considerations from atomic physics or the classical theory of waves. A frequency interpretation was simply unavailable: such data could not be regarded as from a specified population. For Fisher, the central concept was the population, which remained unchanged between repeatable experiments, and which revealed itself imperfectly and partially by the random scatter of specific data sets. For Jeffreys, in contrast, it was the data itself that was fundamental. Consequently, each of Jeffreys's results had to be handled on its own terms. Weighed in the light of experience and expertise, its reliability needed to be reduced to a single number in order that it could be compared and combined with other kinds of evidence. Hence likelihood methods or tools for the analysis of large sequences of data were not appropriate. Instead, Jeffreys sought a single unifying inferential technique that would admit of different kinds of evidence with different reliabilities.

Jeffreys's study of geophysics was based on observation. Likewise his theory of probability, which codified the practices of the scientific community. Fisher's genetics depended on a theoretical model of heredity. The theory of statistics was similarly a construct, which when followed would ensure standard and objective experimental practices. Jeffreys was interested in the truth rather than the utility of science. How did the solar system form? He cheerfully conceded that his geophysical work had no applications at all. Fisher, in contrast, needed to supply agriculturists with a convincing measure of the superiority of one sort of soil preparation over another. Thus even though

---

[119] Writing to Fisher in 1938, Jeffreys commented that "My personal reaction is one of envy of people that can design their experiments with any hope that the arrangement will be uncorrelated with the thing they are trying to find." HJ (Jeffreys to Fisher, May 19, 1938).

Fisher and Jeffreys recognized each other to be promoting a separate form of probability, they were too close to their respective disciplinary experience to appreciate that the choice between the epistemic and frequency interpretations was not merely a matter of cosmetics or convenience, but a consequence of opposed views of scientific aims and practice.

6

Probability During the 1930s

6.1. INTRODUCTION

Jeffreys was not alone in advocating inverse probability. Indeed, Fisher was unusual among British statisticians in attacking it. According to Keynes, use of the Principle of Insufficient Reason was widespread and unquestioned during the 1920s. Keynes's *Treatise on Probability* further persuaded philosophers and scientists of the force of the epistemic interpretation. The criticisms of the method that had surfaced in the nineteenth century had led not to its abandonment but to the recognition that it was limited and must be treated cautiously. Most mathematical textbooks continued to include inverse probability, though they rarely showed applications to real situations. Sandy Zabell quotes a revealing passage in which the Harvard mathematician Julian Coolidge wearily conceded that despite Bayes's defects, it was the "only thing we have to answer certain important questions which do arise in the calculus of probabilities . . . Therefore we use Bayes' formula with a sigh, as the only thing available under the circumstances."[1]

Neither Jeffreys nor Fisher was able to convince the other of his interpretation of probability. Yet the respect, often grudging, accorded the inverse method gradually subsided during the 1930s, and the frequency interpretation – with Fisher's associated approach to scientific experimentation – gained hold. In the previous chapter I argued that each of Fisher's and Jeffreys's methods was coherent and defensible, and that the clash between them was not a consequence of error on one side. What then explains the eclipse of Bayesianism? The answer is found not so much in the conceptual limitations of the theory – which, after all, had long been known – as in the local resources used to interpret probability. These depended on the context of method and practice, and combined in markedly different ways across disciplines. The

---

[1] Coolidge 1925, p. 100, as quoted in Zabell 1989b, p. 254. Zabell writes, p. 252, that the attacks on inverse probability Fisher credited to Boole, Venn, and Chrystal merely caused "the exponents of inverse methods to hedge their claims for the theory."

processes of interpretation involved not only the evaluation of probability models, but their construction too. This chapter sketches the development of probabilistic thought across the disciplines of statistics, the social sciences, the physical and biological sciences, mathematics, and philosophy.

6.2. PROBABILITY IN STATISTICS

6.2.1. The position of the discipline to 1930

Karl Pearson continued to dominate statistics during the first third of the century, and his Galton Laboratory, founded in 1904, became the model of organized research. New methods remained tied to particular applications, but the range of these widened as students from a number of different fields traveled to University College to learn Pearson's techniques. *Biometrika*, published from 1901, still concentrated on biometrical and social studies, but began to include statistical treatments of agriculture, astronomy, and economics. Pearson also applied his methods to therapeutic data. In 1904, his unfavorable assessment of the British Army's anti-typhoid serum was instrumental in the establishment of clinical trials for medical treatments. When the Medical Research Council was formed in 1913, a medical statistics department under the Pearsonian John Brownlee was included, with the first-appointed medical statistician, Major Greenwood, a former Pearson student. The traditional link between statistics and eugenics was further weakened during Word War I as Pearson's group was put to work analyzing ballistic data and other military problems.[2]

University College was not the only site of statistical research. Arthur Bowley, who had lectured weekly on the subject at the London School of Economics from its establishment in 1895, worked during the early years of the century on sampling, and in 1912 conducted a successful survey of poverty in Reading.[3] But Pearson's shadow loomed large. Most British statisticians had trained under him, and his department of applied statistics – grown from the Galton Laboratory in 1911 – was for many years the only place to take a degree in the subject. Other university courses remained tied to particular disciplines. At Cambridge, for example, Yule taught a course to economists and biologists, while Eddington lectured on the combination of errors to astronomers and mathematicians. Students – many of them foreign – started

---

[2] Maunder 1977.

[3] Bowley's book, *Elements of Statistics*, first published in 1901, was a standard text for many years, and his chair at the London School of Economics was the only one in statistics until 1933. Bowley used his study in Reading to call for wage rises for the lowest-paid workers. See Maunder 1977.

to visit Fisher at Rothamsted toward the end of the 1920s, but the dominance of University College as a site for statistical teaching and research was cemented in 1933 when Fisher arrived to take the new chair in eugenics, with Pearson's son Egon as the new professor of statistics. During the 1930s, the subject became common in curricula. Graduates course were offered in Cambridge from 1931, and in the following years other universities including Edinburgh and Leeds introduced specialist courses, often aimed at students in associated agricultural research institutes.

The institutionalization of statistics was speedier in America. E.A. Smith, who had trained in Germany, began teaching the subject in 1915 at Iowa State College; three years later, and against resistance from the university's mathematicians, H.L. Rietz started a course at the University of Iowa.[4] Also in 1918, a department of biostatistics was founded at Johns Hopkins by R. Pearl and L.T. Reed. In 1927, George W. Snedecor and A.E. Brandt established a government-funded statistical laboratory at Iowa State, with the aims of analyzing crop production data and training staff from the Department of Agriculture. A number of new departments sprung up during the 1930s, including those at George Washington University and at the University of Michigan at Ann Arbor.[5] By 1939, John Wishart could note that in America "no university was considered complete" without its department and chair in statistics.

6.2.2. *Mathematical statistics*

Though few of the older generation were expert mathematicians – on which more later – Fisher's work from 1921 provoked a number of the younger statisticians to consider the foundations of their discipline. From the mid 1920s, a steady flow of purely theoretical work began to appear in the journals, much of it coming from the groups at University College and Rothamsted. Starting in 1931, Oscar Irwin published an annual review of such 'recent advances in mathematical statistics' in the *Journal of the Royal Statistical Society*. Irwin had moved from the Galton Laboratory to Rothamsted in 1928 to work with Fisher, and his reviews were initially intended to decode and explicate the Great Man's achievements for a wider statistical audience. But his bibliographies also attest to the growth in mathematical statistics. Fewer than 50 papers on the subject listed in 1931 had grown to 400 a decade later.[6]

---

[4] Rietz published an influential textbook in 1927.
[5] See Harshbarger 1976.
[6] Irwin listed 45 publications in mathematical statistics in 1930, 89 in 1931, 91 in 1932, 117 in 1933, and 133 in 1934. From 1934, finding the field too large to survey alone, he was joined by statisticians including W.G. Cochran and H.O. Hartley, and soon thereafter

The Polish statistician Jerzy Neyman made significant contributions to this literature. Initially teaching statistics at Warsaw University and the Central College of Agriculture, he arrived at the Galton Laboratory on a Polish government fellowship in 1925 to spend a year with Karl Pearson. Neyman soon became friends with Egon Pearson, who had lectured at the Galton since 1921. Egon buffed Neyman's English in three papers on sample moments he published in *Biometrika*, and the two corresponded regularly from the end of 1926, when Neyman headed to Paris for the second year of his fellowship. Though Egon Pearson had been immersed in the biometric tradition from an early age, Fisher's criticisms had stimulated him to re-examine his father's statistical methods. Fisher had emphasized the weaknesses of inverse probability, and by introducing the infinite population had illuminated much about the philosophy of inference. Yet Pearson believed Fisher's theory incomplete. He was dubious of the likelihood ratio, though "straightforward and usable" and "intuitively appealing," and regarded estimation as merely one of many tests that could be validly applied to a set of data.[7] Pearson had corresponded with Gosset on the matter, and soon began to consult Neyman too. Their paper of 1928 discussed five distinct ways to test whether two samples had come from normal populations of the same mean. Of inverse probability, they remarked that the problem of priors was "of no great importance, except in very small samples, where the final conclusions will be drawn in any case with some hesitation,"[8] and went on to suggest that the choice of methods might in the end be "a matter of taste." That said, they "preferred" not to use inverse probability, like Fisher seeking a general method that could be used to test statistical hypotheses irrespective of any *a priori* judgments. Fisher's significance test did not fit the bill: it was concerned only with rejection of the null hypothesis. But this is only half the matter. Neyman and Pearson pointed to an 'error of the second kind': accepting a false null hypothesis rather than rejecting a true one. Using this distinction, Neyman and Pearson introduced a general method for deciding between exhaustive sets of competing hypotheses.

The simplest Neyman–Pearson significance tests assess two hypotheses, one of which must be true. In such cases, the two kinds of errors correspond

limited the article to a bibliography accompanied by the occasional specialist review. He catalogued 586 papers appearing from the end of 1934 to mid 1937, 520 from then until the beginning of 1939, and 381 during 1939, by which time the restricted war-time availability of foreign journals began to have noticeable effect. From 1934, the publications are sub-divided into the following categories: biographical, pure theory, mathematical applications to agriculture, genetics, population and actuarial, psychology, biology, economics, miscellaneous, and, from the 1939 collection, computational.

[7] Pearson 1966, p. 463.
[8] Neyman and Pearson 1928, p. 197.

to accepting the first hypothesis when the second is true, and accepting the second when the first is true. By convention, the more serious of these errors is termed a type I error, the less serious a type II error, and the hypothesis erroneously rejected under a type I error the null hypothesis. The probability of rejecting a true null hypothesis – that is, the probability of a type I error – is usually called the 'size' of the test, and the probability of rejecting a false null hypothesis – that is, one minus the probability of a type II error – the 'power' of the test. As in Fisher's theory of significance testing, an experimental model is defined, and the probabilities of each possible outcome calculated assuming first that the null hypothesis is true, then that it is false. The next step is to select those possible outcomes under which the null hypothesis should be rejected. Though this region of the outcome space would ideally make both the size and power of the test high, increasing one tends in practice to decrease the other. Neyman and Pearson recommended fixing the size of the test – for example, at Fisher's conventional 5% – and then choosing the rejection region to maximize the test's power. The experiment is then performed, and the decision of which hypothesis to accept and which reject is automatic. Neyman and Pearson argued that theirs was a more useful model than Fisher's for many practical situations. Courts of law, for example, are charged with deciding between guilt and innocence, not the probability of guilt. In such cases, the two types of error are also assigned different weights: convicting the innocent is normally considered a greater injustice than acquitting the guilty.

Egon Pearson took over from his father as head of a new department of statistics in 1933. He soon found support for Neyman, who returned to London in the spring of 1934 on a temporary post made permanent the following year. At first, Neyman's relations with Fisher, working in the Galton Laboratory in the same building, were polite. Though Fisher had not been able to help Neyman find an academic position, he proposed him for membership of the International Statistical Institute in May 1934, and several times extended invitations to Rothamsted. They were cordial over academic matters too. Reading a paper on sampling to the Royal Statistical Society in June 1934, Neyman praised fiducial probability as independent of prior distributions; in reply, Fisher described Neyman's account as "luminous."[9] Neyman returned

---

[9] Fisher's praise was not unqualified. He doubted the validity of Neyman's extension of the fiducial method, and disapproved of his reformulation in terms of 'confidence intervals.' Yet Fisher was one of the few in attendance to speak favorably of Neyman's work. Bowley, for example, wondered whether "the 'confidence' is not a 'confidence trick.'" Neyman 1934, pp. 562–3, 589–606; for Fisher's comments see pp. 614–619; for Bowley's see p. 609.

the favor in December following the harsh reception of Fisher's paper on inductive inference. Though continuing to promote his own alternative methods, he hailed Fisher's contributions to mathematical statistics as ground-breaking and damned his critics as old-fashioned.[10]

6.2.3. The Neyman–Fisher dispute

The harmony was short-lived. Fisher learned that Neyman was disparaging his methods to the statistics students; he felt especially resentful having only reluctantly acceded to Egon Pearson's demand the previous year that he not compete with the statistics department by presenting his own ideas on the subject. The final break came in March 1935, when Neyman read a paper to the newly-formed Industrial and Agricultural Research Section of the Royal Statistical Society. Though the subject – agricultural statistics – was very much Fisher's bailiwick, Neyman was sharply critical of Fisher's innovations, questioning both the importance and the unbiased nature of Latin Squares. Fisher's response was characteristic. He "had hoped that Dr. Neyman's paper would be on a subject with which the author was fully acquainted, and on which he could speak with authority," but "since seeing the paper, he had come the conclusion that Dr. Neyman had been somewhat unwise in his choice of topics." But Neyman was equally direct in reply, and an unseemly exchange followed. Fisher was protective of his own work, and charged Neyman with missing its point. ("I suggest that before criticizing previous work it is always wise to give enough study to the subject to understand its purpose. Failing that it is surely quite unusual to claim to understand the purpose of previous work better than its author.") The normally tactful Egon Pearson sided with his bellicose collaborator and flung this accusation straight back: Fisher should not have described Neyman as incompetent without "showing he had succeeded in mastering his argument." Fisher, it was implied, was bamboozled by the equations. Since he had himself called in 1921 for pure mathematicians to follow in his pioneering wake, he was in no position to complain when they discovered he had taken a wrong turn.[11]

---

[10] Fisher 1935b; see pp. 73–76 for Neyman's comments and pp. 81–82 for Fisher's appreciative reply. For the relationship between Fisher and Neyman see Reid 1982, pp. 114–121; also Box 1978, p. 263, who writes that Neyman had an "apparent, if undeliberate, genius for making mischief."

[11] Neyman 1935; see pp. 154–6, 173 for Fisher's comments; for Pearson's comments, see pp. 170–1. That Fisher had just begun his clash with Neyman might account for the unexpectedly cordial tone of his correspondence with fellow-scientist Jeffreys over inverse probability.

Fisher was constitutionally unable to back down from such a confrontation, but in Neyman he had found a formidable opponent. Though lacking Fisher's command of language and polemical skills, Neyman's mathematical genius made his arguments confident and forceful. Zabell shows that direct attacks in print between the two were rare, until the mid 1950s at least, pointing instead to the "poisonous atmosphere" of the common room shared by the two groups.[12] But indirectly the feud boiled. Frank Yates, Fisher's lieutenant at Rothamsted, immediately joined with Neyman over randomization and the properties of the Latin Square, and their exchange continued in the journals for many years. As in Fisher's battle with Pearson, the rhetoric of the dispute settled into a pull between mathematical rigor on the one hand and practical know-how on the other. This time, however, the mathematical amateur was Fisher. Neyman charged his work as inconsistent and careless. Fisher used the tail probability of a test statistic – the total area beyond the observed value – as a measure of the significance of a particular set of results when compared with the null hypothesis. But he could not explain why the tail should be favored over any other region of low probability. It was only under the Neyman–Pearson scheme that the use of the tail probability could be justified. Fisher was unused to his role as practical man. But his footwork was deft. Mathematical abstraction to the neglect of scientific application was useless, he declared. Neyman, with no experience of practical research, had been misled by his "algebraic symbolism."

The debate centered on matters of interpretation, blurred sometimes even for the protagonists. Though by 1938 Neyman sharply distinguished the confidence level and fiducial approaches – and dismissed the latter – he had initially regarded them as similar[13]; for his part, Fisher never regarded confidence levels as genuinely innovative. As another example, consider again the significance level. Fisher claimed that this was a kind of error probability, and that the value observed told on the results of a particular experiment.[14] But Neyman argued that the significance had no interpretation in terms of sampling frequencies, and that such levels could not in any case serve as objective indicators of plausibility because they could not be meaningfully compared between different experiments, null hypotheses, or even test statistics. Neyman believed that nothing could be said about individual experiments: the level of significance actually observed in any particular case is largely

[12] For a summary of the mathematics of the dispute see Zabell 1992, esp. pp. 374–376.
[13] Neyman 1938, pp. 158–9.
[14] See Fisher 1935a, §7; also Fisher 1935b, pp. 48–51. Fisher later admitted this interpretation was problematic.

irrelevant. Instead, the pre-determined type I and type II error rates reflect the long-run behavior of the test procedure, and give the frequencies with which we will reject a true null hypothesis or accept a false one when the experiment is performed many times.

The Fisherian and Neyman–Pearson approaches are mathematically similar, and tend to lead to the same numerical results. Fisher's tail probabilities, for example, can be associated with the probabilities of avoiding errors of Neyman and Pearson's first kind. Such similarities, however, conceal a fundamental difference in philosophy. Fisher's is a theory of inference. It enables the numerical expression of confidence in a hypothesis or theory. The Neyman–Pearson approach, in contrast, governs behavior: it gives a self-contained decision strategy between courses of action.[15] The Neyman–Pearson approach is particularly suited to practical applications, for which the benefits and penalties associated with various well-defined hypotheses can often be quantified. Consider quality control, an active interest of Egon Pearson's since the physicist W.A. Shewhart from the Bell Telephone Laboratory spoke on the subject at University College in 1931.[16] Statistical methods can be used to address the variation inherent in all manufacturing processes, and the relative utilities of the two kinds of errors – corresponding here to rejecting good products and accepting duds – can often be precisely related to production costs and lost profit. For a desired level of significance or stringency of control, the Neyman–Pearson approach can be used to calculate precisely the power and size of sampling test required.[17]

Decisions concern the rational way to behave: what one wants to know is not whether a given hypothesis is true, but whether one should act as if it is. Neyman began to deny the possibility of statistical inference around 1935, talking instead simply of 'inductive behavior.' But in pure science the approach is less appropriate. Fisher strongly believed that a mathematical theory of statistics should have an epistemic interpretation. Science was the search for truth, not efficiency. He argued that the utility of a particular scientific theory should not enter the investigation of whether it is true. And away

---

[15] Neyman and Pearson 1933 concludes that "a statistical test may be regarded as a rule of behavior to be applied repeatedly in our experience when faced with the same set of alternative hypotheses." Pearson 1966, p. 464, suggests that Jeffreys probably refereed this paper.

[16] See Pearson 1933. Shewhart, who had worked on quality control from 1924, had just completed his pioneering book on the subject, *Economic Control of Quality of Manufactured Products*.

[17] Fisher's design prescriptions led to similar results, but in a somewhat less unified and self-contained form.

from manufacturing it was rarely possible or desirable to quantify the two kinds or error in any case. By how much is it better to mistake a genetic factor than to risk delaying publication regarding a possibly remediable disease? The trouble, Fisher said, was that Neyman and Pearson did not appreciate the nature of scientific experimentation. Abstract mathematics could not address real problems. Often, for example, ongoing research would lead a scientist to question the basis of his model. This was a matter not of rejecting one of Neyman–Pearson's competing hypotheses in favor of the other, but both in favor of a different description. Fisher's battery of tests allowed for such a contingency; the Neyman–Pearson scheme did not. Fisher strongly disliked the mechanical nature of the Neyman–Pearson test. Science could not be reduced to pure mathematics. The choice of methods and test statistics required "[c]onstructive imagination, together with much knowledge based on experience of data of the same kind."[18] As the debate progressed, Fisher's language grew to resemble that of Jeffreys. He spoke of 'learning from experience,' the always provisional nature of hypotheses, and the necessity of individual expertise.

During the 1930s, the *Supplement to the Journal of the Royal Statistical Society*, which catered to the more mathematically inclined members of the society, bubbled with ill-concealed animosity.[19] The clash between Fisher

[18] Fisher 1939, p. 6. See also Fisher's concluding remarks to a lecture on the nature of probability delivered at Michigan State University in 1957: "Of course, there is quite a lot of continental influence in favour of regarding probability theory as a self-supporting branch of mathematics, and treating it in the traditionally abstract and, I think, fruitless way. Perhaps that's why statistical science has been comparatively backward in many European countries. Perhaps we were lucky in England in having the whole mass of fallacious rubbish put out of sight until we had time to think about probability in concrete terms and in relation, above all, to the purposes for which we wanted the idea in the natural sciences. I am quite sure it is only personal contact with the business of the improvement of natural knowledge in the natural sciences that is capable to keep straight the thought of mathematically-minded people who have to grope their way through the complex entanglements of error, with which at present they are very much surrounded. I think it's worse in this country than in most, though I may be wrong. Certainly there is grave confusion of thought. We are quite in danger of sending highly trained and intelligent young men out into the world with tables of erroneous numbers under their arms, and with a dense fog in the place where their brains ought to be. In this century, of course, they will be working on guided missiles and advising the medical profession on the control of diseases, and there is no limit to the extent to which they could impede every sort of national effort." Fisher 1958, p. 274.

[19] The *Supplement*, which started publication in 1934, was the organ of the Industrial and Agricultural Research Section of the Society. This had been established in 1933 following the discussion of Pearson's paper on quality control, in which a consensus was reached that industry needed to take more notice of statistics and aim for a standard method of analysis. Members of the new section included Wishart, Hartley, and Bartlett from Cambridge, Pearson from University College, and a number of industrial statisticians,

and Neyman over significance tests, though the most consequential, was just one of a number of fractious spats. Most involved Fisher. In late 1934, he fell out with Wishart, who had continued in print to ascribe to one of his students a proof that Fisher had rejected for publication on the grounds that it was covered by earlier work of his own.[20] In March of 1936, his old dispute with Student over randomization, smoldering for over a decade, flared up and was still ablaze at the time of Student's death in 1937.[21] One of the most far-reaching of these debates concerned what became known as the Behrens–Fisher problem, the estimation of the difference between means of two normal distributions with different variances.[22] Fisher extended his fiducial method to the multiparameter case in his first attempt at the problem in 1935. His results – which Zabell describes as "bold, clever, but in many ways rash" – were queried by Bartlett from 1936 on the grounds that Fisher's calculated significance levels did not square with the sampling frequencies. Bartlett raised additional problems with the fiducial argument from 1939; although his general point was valid, his chosen illustration was not, which considerably prolonged the confusion.[23]

6.2.4. The Royal Statistical Society

F.N. David has remarked that "the seven years 1931–1938 gave the Society a reputation for bitter quarreling which it has not yet entirely cast aside, but . . . was extremely fruitful in that the logic of the interpretation of the data

including W.J. Jennett and B.P. Dudding from the General Electric Company (G.E.C.), L.R. Connor and E.S. Grumell from Imperial Chemical Industries (I.C.I.), E.C. Fieller from Boots Pure Drug Company, H.E. Daniels from the Wool Industries Research Association, and A.T. McKay from the British Boot, Shoe, and Allied Trade Research Association. In 1941, with many young statisticians occupied with war work, the section ceased meeting and the *Supplement* stopped appearing, but late the following year, Dudding helped establish a new Industrial Applications Section, harnessing the war-time interest in statistical control of mass production. See Plackett 1984.

[20] The student was S.S. Wilks, later to found the statistical program at Princeton. See Wishart 1934 and following discussion; Box 1978, p. 267; also the relevant letters in Bennett 1990.

[21] Student criticized randomization on the grounds that with judicious selection, an expert experimenter could obtain results of greater precision. Of this dispute Fisher's biographer, Joan Fisher Box, p. 270, writes: "Fisher's initial rage was excessive, and his later actions lacked magnanimity."

[22] Independent of Fisher, the German agriculturist W.U. Behrens had raised the issue in 1929.

[23] See Bartlett 1965; also letters to Bartlett in Bennett 1990. See also Zabell 1992, pp. 376–379; Barnard 1987, 1995, who attributes the problem to the fact that parameters that give rise to fiducial distributions are not random in Kolmogorov's sense.

collected by the statisticians was finally established."[24] At the time, however, the logic of interpretation seemed to interest few members of the society. Until World War II, it was dominated by economists, and the majority of papers appearing in the journal focused on matters such as the coal supply, the iron and steel industry, and the export trade.[25] The more 'social' studies also tended to have an economic bent, such as papers on unemployment and welfare, the comparison of wages, and worker migration. Most books reviewed were on economic or industrial themes, and tables of trade figures, retail prices, business indices and the like appeared at the back of each issue. Little of this work was mathematical. The tables of data that typically accompanied each article tended to go largely unanalyzed. Authors used statistical methods for data reduction and to illustrate observed trends, but numerical investigations of, say, the relationship between average wages and quality of housing were rare, and the few inferences drawn were qualitative and discussed in the text.[26]

It is thus perhaps not surprising that the society responded to the flow of mathematical papers during the 1930s with positive antagonism. Fisher's eminence in mathematical statistics was soon recognized, but so too was the abstruse nature of his work and its apparent distance from statistical practice. The older members did not hide their antipathy. In 1930, H.E. Soper published a paper purporting to translate Fisher's work on sampling moments for the benefit of the general statistical reader.[27] His polite veneer quickly wore to frank sarcasm. "For more lofty flights into this singularly alluring but secluded region of mathematical statistics Dr Fisher's new moments and product moments and his careful exposition of the theory will be welcomed as a real advance ... If the terrifying overgrowth of algebraical formulation

[24] See Plackett 1984, p. 150.

[25] Four of the five presidents during the decade were connected with financial statistics. Sir Josiah Stamp and Arthur Bowley were economists; Lord Meston and Lord Kennet were government secretaries with responsibilities for trade. (The fifth was Major Greenwood, a medical and epidemiological statistician.)

[26] The society met from November to June, on the third Tuesday of the month. The paper presented at each meeting would appear in the Society's journal, as would an account of the following discussion. Mathematical papers number only half a dozen at most during the 1930s; fewer still had appeared before 1930. See Hill 1984. The focus on the collation of empirical data was typical in other statistical societies too. The International Statistical Institute, for example, remained the preserve of official state statistics through the 1930s. Its congresses were divided into demographic, economic, and social sections, with no special place for mathematics.

[27] Soper, a student of Pearson, had been first author on the 1917 paper attacking Fisher's method of maximum likelihood. He remained wedded to the epistemic interpretation of probability, and died later in 1930.

accompanying this branch of statistical enquiry is destined to have a chief utility in induction and going back to causes, then perhaps Dr Fisher's way of extracting a sample will prove to be most fertile . . . " He concluded with an empurpled attack on the mathematical approach to statistics: "May we not ask, where are the easements in our apprehension of presented phenomena that offset in the least degree the intolerable loads put, in this set itinerary, upon the shoulders of chance eventment? Do we fly better for heaping our wings with micrometer gauges to render exact the survey of the celestial dome that is not even in sight!"[28]

Soper's sentiment was typical, and distrust of small sample work and statistical inference common. The much-respected Major Greenwood worried that practicing statisticians were dazed into inaction by statistical analysis, or worse, left in thrall to the mathematicians.[29] Though a visiting industrialist might from time to time be chided for his lack of statistical sophistication, in general "learned" and "theoretical" contributions were respectfully dismissed. Good old common sense was of more value to the practical man. Discussing Pearson's 1933 paper on quality control, B.P. Dudding drew attention to the "pitfalls for the pure statistician who plunged himself into a factory and thought he could put the manufacturer right even inside a period of ten years . . . in industry one always got driven back finally to the use of a little common sense."[30] As the industrial statistician L.R. Connor put it: "Guessing a solution and working backward may lack theoretical polish, but it frequently leads to practical results in less time than would have been consumed by more elaborate processes."[31] The suspicion of mathematical statistics was reflected in attendance at the society's meetings. One speaker noted that if a late-arriving fellow "should experience difficulty in finding a comfortable seat there is no doubt whatever . . . that the intricacies of finance are under discussion; if, on the other hand, he observes a somewhat sparsely filled hall it is more than likely that the statistics at issue are those of life and death; while if he is faced with an embarrassingly empty room it is certain that an exponent of mathematical methods is at the reader's desk."[32]

---

[28] Soper 1930, pp. 110–1, p. 114.

[29] See Greenwood 1932, particularly pp. 721–2.

[30] Dudding's comments in Pearson 1933, pp. 66–8. See also the discussions following A.P.L. Gordon's 1931 paper on the statistics of totalisator betting, and E.C. Rhodes's 1937 paper on a business activity index. Both Gordon and Rhodes regarded their critics as overly-mathematical hair-splitters.

[31] Connor 1931, pp. 442–3.

[32] Hill 1936, p. 162. The speaker, Austin Bradford Hill, later worked under Fisher to persuade the medical profession of the efficacy of randomized clinical trials.

6.2.5. The reading of Fisher's 1935 paper

The establishment in 1933 of the Industrial and Agricultural Research Section eased some of the tension, but by concealing rather than resolving differences in approach. The mutual antipathy is exemplified by the hostile reception given to Fisher's 1935 paper on the logic of inductive inference. True, in presenting a highly condensed account of his work, with "a few novelties which might make the evening more interesting to those few among the audience who were already familiar with the general ideas," Fisher did not help matters. (Barnard writes that this was a "wild overestimation of the capacity of his audience."[33]) Even so, the response was discourteous. Proposing the vote of thanks, Bowley referred to the paper's obscurity before attacking it as arbitrary and misleading,[34] then suggesting Fisher had insufficiently credited Edgeworth for his ideas. Leon Isserlis continued the attack, despite admitting his bafflement. He doubted the novelty and applicability of Fisher's work, and saw no good reason to reject inverse probability. ("A good deal of work has been done in showing that in the cases that matter, the extended form of Bayes' theorem will suggest as a reasonable estimate very much the same value as that suggested by the method of maximum likelihood."[35]) Professor Wolf, a philosopher brought in by Bowley as a hired gun to pick at Fisher's logic, agreed that Fisher "had proposed a very ingenious method of making a little evidence go a very long way," but declared that he couldn't decide whether Fisher's measures of reliability were subjective, or objective and based on frequencies, and thus properly scientific in character. He then embarked on a condescending tutorial on the philosophical basis of uncertain inference. Greenwood was too mannerly for direct criticism, but suggested that Fisher might try presenting his work in a more comprehensible way. In reply, Fisher was contemptuous. "I find that Professor Bowley is offended with me for 'introducing misleading ideas.' He does not, however, find it necessary to

---

[33] Barnard 1990, p. 27.

[34] "I took it as a week-end problem, and first tried it as an acrostic, but I found that I could not satisfy all the 'lights.' I tried it then as a cross-word puzzle, but I have not the facility of Sir Josiah Stamp for solving such conundrums. Next I took it as an anagram, remembering that Hooke stated his law of elasticity in that form, but when I found that there were only two vowels to eleven consonants, some of which were Greek capitals, I came to the conclusion that it might be Polish or Russian, and therefore best left to Dr Neyman or Dr Isserlis. Finally, I thought it might be a cypher, and after a great deal of investigation decided that Professor Fisher had hidden the key in former papers, as is his custom, and I gave it up." Bowley's comment in Fisher 1935b, p. 56.

[35] Isserlis's comment in Fisher 1935b, p. 59. Fisher replied that his results were conceptually different from Edgeworth's, which were contaminated by his "deep entanglement" with inverse probability.

demonstrate that any such idea is, in fact, misleading. It must be inferred that my real crime, in the eyes of his academic eminence, must be that of 'introducing ideas.'"[36]

The problem was not merely the mathematical naiveté of the older members of the society, rather that their professional aims conflicted with Fisher's approach. This is why he had not previously been invited to read to the society, despite it's being thirteen years since the publication of his first major works, and despite *Statistical Methods for Research Workers* having reached a fifth edition. Most members were chiefly concerned with establishing themselves as links between government departments, business, and the public, and interpreting for each the meaning and significance of officially-gathered figures. Thus they pressed for statisticians to form committees to advise the government (to play the role of "confidential waiting-maid to the princess,"[37] as Greenwood put it), for the scope of official statistics to be extended, and for all statistics to be gathered and published more rapidly. The Manchester Statistical Society, also business-oriented, enlisted the support of the London Society in a memorandum on the subject to the Minster of Labour. Though the response was gently non-committal, many in the society continued to lobby for statisticians to take the role as impartial gate-keepers of official statistics.[38] In his inaugural address as President in 1937, Lord Kennet argued that statistics were necessary to keep the public safely informed and to allow the legislature to make informed decisions. He suggested that the Society was "a distinguished middleman ... an honest broker between the consumers of statistics and the producer. It tries to help the producer to put

---

[36] Fisher 1935b, p. 77. The more mathematical members did offer some words of support. Egon Pearson defended Fisher's contributions in reconstructing the logical foundation of inference; in written communications Bartlett too was broadly appreciative, and Neyman – who had not yet broken with Fisher – hailed Fisher's fundamental contributions to mathematical statistics, and sympathized with his being subject to criticism from those too wedded to the old views. Both Pearson and Neyman did, however, add to the critical atmosphere by taking the opportunity to raise a number of technical issues in Fisher's theory of mathematical inference, and to promote their own methods instead. Jeffreys continued to harp on Fisher's frequency definition of probability, commenting in a written contribution that Fisher's results could better be justified by inverse probability. Fiducial probability "shows insufficient respect to the observed data and does not answer the right question."

[37] See Kennet 1937, p. 14.

[38] The "Memorial on earnings and the cost of living," signed by H.W. Macrosty, Joint Honorary Secretary, on behalf of the President and Council of the Royal Statistical Society, was sent to the Minster of Labour in February 1936. His secretary, T.W. Phillips, replied in standard civil service language that inquiries were being made with a view to acquiring more information, etc. The committee of the Society followed by expressing the hope that the Minister would soon reexamine the matter, and offered help if needed. See Macrosty 1936.

184

sound goods on the market, and to help the consumer to tell between goods sound and unsound."[39] Such 'goods' included matters of social policy. When in 1937 a letter to *The Times* sparked popular debate on the subject of the falling birth-rate, members responded with the demand for a new statistical survey. This was the only way to gather unbiased evidence and thus ensure an appropriate policy could be framed.[40]

Most of these efforts were frustrated. Sir Josiah Stamp lamented seeing "Cabinet Ministers accepting our statistics before they go into a Cabinet meeting, and emerging having forgotten all about them," and complained that without access to statistics the public were being hoodwinked by scaremongering in press.[41] The issue was one of trust, something I shall say more about in relation to the social sciences. In short, though, the statisticians found that to stand a chance of influencing government policy, their recommendations must not be seen to be tainted by opinion or interest. In the words of the memorandum, statistical investigations must be organized and conducted so to be "free of bias and tendentiousness." Major Greenwood put the matter clearly in a well-received response to an analysis of housing in Liverpool: official statistics should be protected against "zealous social reformers. Frankly, the suggestion that official tabulations should be influenced in any way by considerations of what is socially desirable is a thing we statisticians ought to resist to the death. Directly one turns from the recording of alleged facts to anything approaching propaganda, that is the end of impartial official statistics. More than one hopeful venture in other countries has thus come to shipwreck. The surpassing merit of the Registrar-General's statistics has always been that they are objective."[42]

Now Fisher regarded his work as perfectly objective; it was precisely the inverse method used by the *ancien regime* that was subjective and thus questionable. But most of the older statisticians were not concerned with such mathematical minutiae. Since they tended to interpret probability epistemically, they rejected *any* form of inference as too close to speculation. The economist R. Glenday pointed to the pitfalls associated with fitting "the complex

---

[39] Kennet 1937, p. 1. See also p. 7: "It is a very sober truth that under modern conditions at home and abroad with the State invading the whole sphere of human life, humanity cannot get on without the statisticians, to tell it how things stand, and what the effects of its political actions will be."

[40] The letter had been written by Sir James Marchant. See Greenwood 1937 and following discussion. In one of his few contributions to discussions of the main society, Fisher predictably called for greater emphasis to be placed on the quality rather than the quantity of children.

[41] See Kennet 1937, pp. 16–17.

[42] See Jones and Clark 1930, p. 522.

confusion of human events of the economic world into a frame of formal thought derived *a priori*," and called for the gathering of statistics to be clearly separated from any attempts at inference.[43] Karl Pearson's influence was clear. The majority of members took the attitude that statistics were useful just for summarizing data and finding correlations. They were suspicious of Fisher's fancy mathematics, and his attempts at a theory of statistical inference.

This attitude persisted until the end of decade. When in 1939 Wishart complained that the close connection of statistics to practical applications worked to the detriment of its purely mathematical development, Bowley pointed to the motto of the society – 'aliis exterendum' (to be threshed out by others) – and thundered that inference from samples to the whole population was "dangerously near the expression of opinion which the Founders of the Society forbade." Bowley's own views were clear. He modeled his 'econometrics' explicitly on biometrics as a study of data reduction and description.[44] In discussing Wishart's paper he was said to be "fully appreciative of the importance of [the Industrial and Agricultural Research Section], and he was glad of its existence, if only as a safety-valve for the people who had to think of statistics in that kind of way."[45] Greenwood also continued to insist that common sense was more important for statisticians than complex mathematics. His sentiment was echoed by V. Selwyn, who had spent three years as a business statistician, and who argued that "no amount of education could give the student the intuition necessary for drawing conclusions. There was a sort of common sense without which no student could be trusted to tackle any statistical problem or draw any conclusion."[46]

---

[43] Glenday 1935, p. 497 inveighed against the obsolete, nineteenth-century assumptions that persisted in his discipline, and spoke with distaste of economists who had "not hesitated, under the stress of circumstance, to use [their purely] formal reasoning as a support for practical proposals of a most drastic and far-reaching kind, without any statistical background whatsoever."

[44] Having written in 1924 the first book on the subject, Bowley can be considered one of the founders of mathematical economics.

[45] Wishart 1939, p. 561.

[46] See Wishart 1939, p. 559. Wishart, by contrast, was concerned with the state of statistics teaching. (I note John Bibby's observation that "groups most interested in teaching are often those most interested in the status of the profession." Bibby 1986, p. 7.) He deplored the quality of mathematics in most statistical studies, and noted that statistics was not even taught in a consistent way across different disciplines. Wishart called for a standard treatise in mathematical statistics, perhaps collectively-authored, to take students beyond Yule and Kendall's introductory text. Optimistically, he predicted that although the topic was controversial, a consensus might be reached on the interpretation of probability to be used. Fisher contributed to this debate by steering a careful line between the mathematical approach and the methodism of Neyman. He noted, p. 554, "the danger that the very word

186

6.2.6. Statisticians and inverse probability

How, then, did statisticians interpret probability in the 1930s? The answer must take account of the differences between practical and mathematical statisticians. In 1939, Jeffreys wrote of Karl Pearson that "though he always maintained the principle of inverse probability . . . he seldom used it in actual applications, and usually presented his results in a form that appears to identify a probability with a frequency."[47] The practical men of the Royal Statistical Society tended to share Pearson's ambiguities.[48] Statistics and probability were thought to be quite distinct: statistics was the numerical treatment of large quantities of data; probability was an epistemic concept that governed inference. Certainly, the inference from a sample to a statement about the whole population required the theory of probability. But since large samples were thought to be the proper object of statistics, statisticians had no need for inference. Inverse probability was therefore rarely useful as a method of statistical practice.

Some statisticians were insulated from resolving the two forms of probability by the assumptions underlying their particular field of inquiry. Economic statisticians generally took agents to be rational, for example, and thus found it easy to aggregate individual behavior and so equate sample with population statistics. Bowley had not used probability in his research because "in descriptive economics the field of observation is not generally regarded as a sample from an infinite world, but as complete in itself."[49] When the old guard did venture into inferential terrain, they were content to travel with inverse probability. Accused by Neyman of basing his inferences on

'mathematical' might come to acquire a certain derogatory import through association with futile logic-chopping on questions on which all competent practitioners had already made up their minds."

[47] Jeffreys [1948], p. 355.

[48] As late as 1939, Wishart noted that "to-day it may even be found that no two authorities agree in all details in what they would regard as a satisfactory probability basis for statistical theory." See Wishart 1939, p. 543.

[49] Bowley's comment in Bartlett 1940, p. 23. In reply to Wishart's paper of 1939, Bowley dismissed the confusion over probability as a narrow mathematical matter, unimportant in both practice and the teaching of statistics, p. 561: "He himself wanted to get rid of [probability], or else to have its basis much more definitely established. It might be well to give it another term, but he did not think one ought to introduce ideas of probability in the teaching of statistics except to those who had definite mathematical ability or had had mathematical training." Sir William Elderton took a similar line, p. 554: "He was not sure that when they got down to statistics it would not be wiser whenever possible to avoid the use of 'probability' and to use rather the term 'proportional frequency' or some other phrase suitable to the particular point being investigated. Perhaps they had suffered in using the word 'probability' from the mathematicians who were mathematicians only."

prior assumptions, Bowley replied with supreme indifference: "I do not say that we are making crude judgments that everything is equal throughout the possible range, but I think we are making some [such] assumption or we have not got any further."[50] Leon Isserlis too thought inverse probability, when suitably limited, was "perfectly legitimate."[51] He used the epistemic interpretation in a study that purported to eliminate the arbitrariness of prior probabilities from inferential statements (by retreating from probabilities to probabilities of probabilities).[52] Keynes, on the council of the society, and lauded as economist, was also treated most respectfully as a statistician; there was certainly no disparagement of his work on probability, even Fisher was deferential.[53]

The practical men therefore did not reject one interpretation of probability in favor of the other; rather they turned their backs on the whole concept, in either guise. Mathematical statisticians, in contrast, were careful to distinguish the two. However, since both the Neyman–Pearson and Fisherian approaches were based on the frequency interpretation, the general result of the debate was a hardening of the opposition to inverse probability. The Neyman–Pearson approach was by 1939 considered the more rigorous of the two by most mathematical statisticians, and is in fact more resolutely frequentist. Though Fisher's frequencies were usually relative to hypothetical infinite populations, those of Neyman–Pearson, although often in practice also unrealized, were based on the more tangible notion of repeated sampling from the statistical model in question; Neyman interpreted his significance levels directly as the frequency with which a correct hypothesis could be expected to be rejected in repeated sampling from a fixed population.[54]

---

[50] See Neyman 1934, p. 609.
[51] Isserlis was puzzled by Neyman's confidence intervals, thinking them nothing more than a restatement of inverse probability. See Neyman 1934, pp. 612–4, quote from p. 613.
[52] Isserlis's 'objective' version of the Rule of Succession states that if "an event has happened once, *the probability, that its probability is greater than 1/3, is greater than 1/2,* or the betting is more than evens that the odds against the event are shorter than 2-1." The frequentist Wisniewski objected on the grounds that the "magnitude . . . interpreted by Mr Isserlis as the probability of certain values . . . does not posses the property of being the 'stochastic limit' of the ratio of cases conforming to a given hypothesis . . . ," and concluded that Fisher's pessimism about attaching probabilities to parameters in a population was regrettably well-founded. See Isserlis 1936; Wisniewski 1937.
[53] See, for example, Irwin 1936, p. 728. For Fisher's deference see Conniffe 1992.
[54] Fisher described fiducial limits in almost identical terms in his 1930 paper, but was obliged to reject the interpretation during his exchange with Bartlett, in part because it is untenable for his solution to the Behrens–Fisher problem. This enabled Neyman – who didn't believe that likelihood could represent a degree of credibility – to castigate Fisher's refusal to interpret significance levels in the same concrete fashion as a sort of Bayesianism.

The rejection of inverse methods, however, was not complete even among mathematical statisticians. Irwin gave a relatively impartial summary of the Fisher–Jeffreys dispute in his 1934 "Recent advances in mathematical statistics" article, for example, and though giving Fisher the last word did not dismiss Jeffreys's position, suggesting instead that it threw "considerable light on the principles of statistical inference." Irwin later published a respectful review of Jeffreys's 1939 book *Theory of Probability*.[55] Similarly, W.G. Cochran – also a Rothamsted alumnus – commented favorably on a book by the Belgian Van Deuren, though it took a similar line on probability to Jeffreys.[56] Neither did M.G. Kendall – by this time assisting Yule with new editions of his best-selling statistics text – dismiss the intuitive approach of Keynes and Jeffreys; indeed, though recognizing that most statisticians adopted the frequentist interpretation he used the 'undefined concept' approach in an analysis of random numbers.[57]

Yet the majority of mathematical statisticians bowed to the frequency interpretation. Fisher and Neyman were authoritative, their mathematical methods were powerful and original, and their nomenclature – 'unbiased,' 'fiducial,' 'efficient,' 'sufficient,' and so on – ringingly objective. The mathematicians were successful in transporting their methods abroad. Fisher lectured at Iowa in 1931 and 1936; Neyman followed with a lecture series of his own in 1937, the next year moving permanently to the mathematical department at Berkeley, and soon establishing a new statistical laboratory.[58]

With slight variations of emphasis, all of the important texts adopted the frequency interpretation. By the end of the 1930s, these included the second edition of L.H.C. Tippett's *The Methods of Statistics* of 1937, which was modeled closely on Fisher's work; the eleventh edition of G.U. Yule's *Introduction to the Theory of Statistics* of 1937; H. Cramér's *Random Variables and Probability Distributions* of 1937; the third edition of L.R. Connor's *Statistics in Theory and Practice* of 1938; as well as C.H. Goulden's *Methods of Statistical Analysis* of 1939, based on Fisher and popular in America;

---

[55] Irwin 1934, p. 138; Irwin 1941.

[56] Cochran 1936.

[57] Kendall and Smith 1938; Kendall 1939. A similar ambivalence was evident in the American Statistical Society. Boole's Challenge Problem concerns the probability of an event given the probabilities of possible causes, and is only meaningful under the epistemic interpretation. E.B. Wilson regarded the problem as no more than a curiosity because indeterminate, but noted that many practical investigations result in incomplete data too, often as a result of bad planning, and that "it then becomes necessary to do the best one can with the data available." Wilson 1934, p. 304. (See Dale 1991, p. 313 for Boole's Challenge Problem.)

[58] Cochran took up a permanent post in Iowa in 1939 to teach mathematical statistics.

and important continental works such as M. Fréchet's *Généralités sur les Probabilités: Variables Aléatoires* of 1937. Throughout, Fisher's *Statistical Methods for Research Workers*, in a seventh edition in 1938, continued to be the most influential source for new students to the field. When in Chicago in the late 1930s, L.J. ("Jimmie") Savage had asked W. Allen Wallis how he could learn statistics, the answer was that he should read and reread *Statistical Methods for Research Workers*. "This is what almost all of that generation of mathematical statisticians did."[59]

The mathematical statisticians and their firmly frequentist views came to dominate the discipline, old members like Bowley notwithstanding.[60] Yule had early seen this coming. At Karl Pearson's death in 1936, he wrote: "I feel as though the Karlovingian era has come to an end, and the Piscatorial era which succeeds it is one in which I can play no part"; in a still more valedictory tone three years later he commented to Wishart on his depression at "how different your statistical world is from mine," before making a last plea for the needs of the non-mathematical economists and psychologists not to be ignored.[61] Bartlett, now at "I.C.I.'s Rothamsted" at Jealott's Hill after a stint under Egon Pearson at University College, was moved to summarize the position in 1940.[62] In a view similar to that expressed during the Fisher–Jeffreys debate, he demarcated mathematical statistics, first from the classical theory of probability and its offshoots in error theory and contemporary mathematical probability, and then from empirical work in social statistics, biology, and psychology. Bartlett defined mathematical statistics as concerned with those processes determined from experience to fit closely with Fisher's model of random sampling from an infinite hypothetical population.[63] Statistical theory is thus just the mathematical approach to observational data arising by chance, or that can be regarded as arising by chance, and developments from

---

[59] As told to George Barnard (GAB, 23 February 1997). Savage [1972], p. 275, said that *Statistical Methods for Research Workers* "had far more influence on the development of statistics in the current century than any other publication," with *Design of Experiments* second.

[60] Maunder 1977, pp. 475–6 notes that Bowley's attitude towards new methods was "unhelpful and negative" towards the end of his career.

[61] Kendall 1952, p. 157; Yule's comments in Wishart 1939, p. 564: "But I would question whether you do not too much ignore the non-mathematician. Most economists and most psychologists are still non-mathematicians. In economic statistics, there is a great deal to be done with little but the simplest of algebra; the non-mathematician may be quite keen, intelligent, and useful, and his needs should not be ignored."

[62] Bartlett 1940, footnote to p. 5; see too Bartlett 1982.

[63] Or more exactly, with a formalized version of Fisher's model due to the Swedish statistician Harald Cramér.

this model constitute the subject of mathematical statistics.[64] This includes statistical inference as a restricted subset. What then of probability? The epistemic interpretation would be significant if degrees of belief were measurable. But they are intangible. If anything, they are to be investigated with statistics, which must therefore be more fundamental. As Bartlett made clear, "there is no place for such topics as inverse probability, if by this term it is implied that the statistician, in contrast with any other scientist, should be obliged to assess *numerical* 'probabilities' of his hypotheses."[65]

6.3. PROBABILITY IN THE SOCIAL SCIENCES

6.3.1. Statistics in the social sciences

The terms 'objective' and 'subjective' are used so indiscriminately that they often seem intended simply to lend a sort of talismanic power to a position or to ridicule that of an opponent. Yet a frequency is more palpable than a degree of belief, and Fisher in his debate with Jeffreys was quick to exploit the force of the language of objectivity to laud his own methods and damn those of the few remaining inverse probabilists.[66] This rhetoric was largely wasted on Jeffreys, for whom issues of standardization and objectivity were not so charged. Working alone, and in an unfashionable field with few applications or implications for policy, there was less need for his results to be transparent. Indeed, while the need for objective statistics was patent to Fisher, the need to account for 'subjective' factors such as the individual investigator's expertise and experience was equally persuasive for Jeffreys. But Fisher's words were anyhow intended for more general consumption. As mentioned earlier, professional statisticians stressed the objectivity of their methods when attempting to convince government agencies that statistical recommendations were trustworthy. Fisher made clear that the objectivity of

---

[64] Bartlett 1940, pp. 13–18 mentioned statistical physics as a good application of the theory. He thought the theory controversial in areas such as psychology and economics, where its assumptions are dubious.

[65] Bartlett 1940, pp. 11–12. In the following discussion, Irwin agreed with this assessment, as did Wishart, who additionally pointed out that "the statistician disputes the premises of [Jeffreys's] opposing viewpoint from the very outset." Kendall too was in broad agreement (though he continued to regard the choice of theory as a "matter of taste"). Neyman had put forward a similar point of view to Bartlett's in 1937; see Neyman 1938, p. 32.

[66] Fisher's commitment to objectivity was strong. To Student's criticism of randomization – more precise results would come from a careful choice of experimental conditions – he responded that the validity and objectivity of the error estimates was worth the price in accuracy.

statistics had a still wider appeal. In *Statistical Methods for Research Workers*, he advertised his methods to the insecure social sciences as a route to credibility: "Statistical methods are essential to social studies, and it is principally by the aid of such methods that these studies may be raised to the rank of sciences."[67]

Fisher and many of the older generation of the biometric school were particularly concerned with the credibility of the science of eugenics, by 1930 seen as the province of romantics and idealists. (Fisher had lost none of his youthful fervor for eugenics, discussing how his theory should be applied to human populations in much of his *The Genetical Theory of Natural Selection* of 1930; he especially lamented that social ability was inversely correlated to family size, and proposed radical schemes of family allowances to encourage the middle classes to breed.[68]) However, social scientists responded enthusiastically to the prospect of an objective mathematical method for experimental design and data analysis. This is not surprising. In the early decades of the century the social sciences were generally distrusted. Sociology, for example, was in the popular mind coupled with ethical or religious interests, and dismissed as a pseudo-science. In America, the confusion with socialism generated additional suspicion. Sociologists of the time sought to counter the opprobrium by reconstructing the discipline as a positive science, to be characterized by quantitative empirical research. Data would be scientific – that is, precise and repeatable – and untainted by prejudice. Sociologists welcomed the firm and universal methodology provided by statistics, and adopted statistical analysis as a substitute for direct experimentation. As Deborah Ross had written, for social scientists "statistics became the visible hallmark of science."[69]

---

[67] Fisher 1925b, p. 2.

[68] Both female emancipation and the increase in contraception, being largely middle-class phenomena, tended to dilute the valuable genetic stock still further. Fisher's solution was a more flexible society, in which the most able were given the best opportunities. Thus gifted children should be well educated, regardless of social class. MacKenzie 1981, esp. p. 188, makes the case that throughout his career it was eugenics that motivated Fisher's interest in heredity. Eugenic considerations certainly spurred his work on the Mendel–Darwin synthesis. Moreover, his sensitivity to the credibility of eugenics made Fisher distrustful of amateurs. In the early 1930s he alienated many in the Eugenics Society by insisting it be led by professional scientists.

[69] Ross 1991, p. 429. See also Gigerenzer 1987; 1989. Porter 1995 is a general account of the relationship between credibility and quantification. He argues that since the appeal to numbers detracts from individual expertise, it is weak communities that tend to quantify, and not from internal pressures but in response to external attacks. Bureaucrats, who have no authority of their own, are an example. Daston and Galison 1992 have also examined the issue of objectivity. In the sense of the removal of the individual, they argue that objectivity

192

A distinction must be drawn here between theory-building and data analysis. In their attempts to win authority for the discipline, social scientists claimed not just the method of the physical sciences but the associated promise of certain knowledge. The laws of society that would emerge from empirical study, they declared, would be as immutable as Newton's law of gravity. With this pretense of determinism, there was no place for chance in their theories. Economists, for example, worked with idealized models based on equilibrium, independence, and perfect competition. Prominent economists including Jevons, Edgeworth, and Keynes had of course written on probability, but each took the epistemic view. And since economic models tended to assume rational agents and a perfect flow of knowledge, none attempted to incorporate the concept into their theories.[70] Edgeworth's 1910 paper, "Applications of probabilities to economics," is not a call for an indeterminate economics; rather it is an argument from the Principle of Insufficient Reason against the assumption of extreme values for economic coefficients when data is incomplete.[71] Economics was largely non-numerical through the 1930s, and even the mathematical economists tended to restrict themselves to unsophisticated tasks such as the use of least squares and correlation techniques to analyze the business cycle.[72] The variation of real data due to factors outside their models – 'external shocks' – was so great that econometricians such as Bowley insisted this work be restricted to description only. They thus rejected rather than just ignored probability theory. Instead, statistics was to be used to extract the relevant parameters of deterministic underlying laws, the forms of which were suggested by theory. Despite the work of a few economists, such as Haavelmo, Koopmans, and Tinbergen, who started to looked to Fisher's inferential methods to test rather than just quantify economic models, this remained the general approach through the 1930s.[73]

The situation was similar in psychology. During the mid-nineteenth century, pioneering experimental psychologists such as Wilhelm Wundt and Gustav Fechner – the 'father of psychophysics' – had shifted the focus of

arose in the early and mid-nineteenth century, and was associated with mechanization and the move from ideal to naturalized depiction. Daston and Galison suggest that it was from the start a moral matter too, tied to the Victorian virtue of selflessness.

[70] Keynes confided in correspondence that the concept of probability was useless for a study of economics. He was even suspicious of econometrics. See Ménard 1987, p. 143.

[71] Edgeworth 1910, p. 287: "There is required, I think, in a case of this sort, in order to override the *a priori* probability, either very definite specific evidence, or the consensus of high authorities." Edgeworth was arguing against a certain Professor Seligman.

[72] See Horváth 1987.

[73] Morgan 1987 argues that economists have only recently started to investigate non-deterministic models and irrational behavior. See also Menard 1987.

the discipline from providing accounts of rational belief to performing experimental investigations of perception. They employed simple statistical techniques to probe beneath the variation of individual experimental subjects. A frequency interpretation grounded this work, since the categories of perception were defined on a scale marked by the ratio of an individual's repeated assessments of the physical stimulus under consideration. But the core theory was still predicated on mechanistic laws. Perceptual variation was taken to be a result of determinate, if unknown, perturbations around a 'true' value of mental experience.[74] Gigerenzer has argued that 'probabilistic thinking,' in the form of Fisher's methods, provided a route toward freeing the subject from opinion and crankery. So while psychologists proved more willing than economists to turn to statistical inference, they were equally leery of admitting chance effects to their theories. Despite ample commonsense experience to the contrary, psychologists of the 1930s such as Kurt Lewin and Clark Hull continued to regard causal laws of the human mind as the holy grail.[75]

6.3.2. Statistics reformed for the social sciences

The rise of quantitative methods in the social sciences during the 1920s, promoted in America by such new bodies as the Social Science Research Council, was followed during the 1930s by an increased use of statistical models for analysis.[76] Initially this 'statisticism' had an institutional flavor. F.S. Chapin and Franklin Giddings led the way at Columbia. Giddings had earlier taken an interest in evolution, and came to statistics through the work of Karl Pearson. He introduced biometrical techniques, such as correlation and goodness-of-fit tests; these gradually gave way to Fisherian hypothesis

---

[74] Fechner was a rare indeterminist among nineteenth-century scientists. He wrote on statistics in his posthumously-published *Kollektivmasslehre* of 1897. See Stigler 1986, Chapter 7; Danziger 1987; Gigerenzer 1987.

[75] Such views persisted into the 1960s and beyond. Edwin Boring sought deterministic laws of psychic states, as did David Krech, who in 1955 beseeched the readership of *Psychological Review*: "I have faith that despite our repeated and inglorious failures we will someday come to a theory which is able to give a consistent and complete description of reality. But in the meantime, I repeat, you must bear with us." These words are quoted by Gigerenzer 1987, p. 13, who argues that the pervasive ethos of determinism retarded B.F. Skinner's full expression of behaviorism. He concludes, p. 16: "Probabilistic thinking seldom threatened psychological determinism. If it did, as with Egon Brunswik, who postulated irreducible uncertainty in the subject matter itself, such attempts were suppressed or forgotten with the understanding that this was not the psychologist's way toward an acceptable science."

[76] See Ross 1991, pp. 390–470, who notes on p. 429 that by 1926, 31 percent of published work in sociology involved statistics. Oberschall 1987 is an account of the introduction of statistics and quantitative methods in French sociology.

testing, coupled with a growing reliance on sample surveys.[77] In late 1934, Fisher's analysis of variance was introduced in experimental psychology to pinpoint causal relationships between variables; from around 1940 it was ubiquitous.[78]

The model of statistics adopted in the social sciences was broadly Fisherian. *Statistical Methods for Research Workers* was aimed explicitly at biologists and human scientists, and sold well in America from 1927 after a favorable review by Harold Hotelling.[79] Fisher's emphasis on small samples, as Hotelling pointed out, was attractive to practitioners often dealing in restricted data sets. Also, his tables were carefully tailored to appeal to researchers in the softer sciences.[80] (Fisher presented the one-sided version of Student's test since it was easier to calculate, though he was well aware that the two-sided version was more robust against skewness.) Fisher's *Design of Experiments* of 1935 proved popular too. Much additional influence was mediated through articles by Hotelling, texts such as Snedecor's 1937 *Statistical Methods*, and from the late 1930s a host of books with titles like "Statistics for Sociologists," "Statistics for Educational Psychologists," and "Statistics for Economists." Writers of these texts were anxious to present a method "cleansed of the smell of agricultural examples such as the weight of pigs, soil fertility, and the effect of manure on the growth of potatoes," and illustrated instead with the peculiarities of their own disciplines.[81]

Though the rapid colonization of social sciences by Fisherian statistics – and the dominance of the associated frequentist interpretation of probability – has been ascribed to the publication of Fisher's books,[82] their enthusiastic reception is as much an explanandum as an explanation. That

---

[77] See Converse 1987. Bowley started publishing on sample surveys from 1906, and continued to develop the method almost single-handed through the 1910s and early 1920s. His 'purposive' method – i.e., the careful selection of a representative group by the experimenter following a rough but complete survey – gave way during the 1920s to random sampling. This was given final theoretical justification by Neyman's paper of 1934. Government bureaux started to rely on sampling methods around the same time. Though official statisticians had long distrusted incomplete surveys, which they held to be speculative and against the descriptive spirit of statistics, the market crash of 1929 showed the practical necessity of fast and cheap assessments of unemployment and other economic indicators. See Hansen and Madow 1976.

[78] See Gigerenzer 1987, Danziger 1987. Sterling 1959 reports that by 1955, over 80 percent of publications in the main psychology journals used significance tests.

[79] Hotelling 1927. Hotelling spent time with Fisher at Rothamsted in 1929.

[80] In 1938 Fisher published with Frank Yates the widely-used *Statistical Tables for Biological, Agricultural, and Medical Research*.

[81] Gigerenzer 1989, p. 207.

[82] For example, Zabell 1989b, p. 261; Good 1980; Barnard 1989, 1990.

Jeffreys's proposals for the analysis of data using the epistemic interpretation were ignored is not surprising.[83] With its degrees-of-belief and arbitrary prior probabilities, the Bayesian view was unlikely to yield results that would be accepted as clear and legitimate knowledge. Also, though both men developed their statistical and probabilistic techniques in response to specific scientific problems,[84] Jeffreys's methods required insight and mathematical dexterity for application elsewhere, while Fisher's led to generalized recipes that could be freely applied in other areas. Further, Jeffreys tended to ignore the problems of data acquisition, while Fisher considered experimental design in detail, especially in his *Design of Experiments*. But the position with the Neyman–Pearson scheme was more complex. Certainly, Fisher's emphasis on inference was congenial for social scientists, who wanted absolute measures of their hypotheses' credibility rather than a guide to behavior when a choice between two was required. But Neyman's influence was growing among statisticians, and his objective approach – more stark even than Fisher's – might have been calculated to appeal to social scientists.

In fact, while statistics made headway as a method of data analysis, social scientists tended to downplay the doctrinal bickering between Fisher and the Neyman–Pearson camp.[85] The natural reluctance of social scientists to concern themselves with the minutiae of interpretation was a factor in this bowdlerization, one justified by the fact that the two methods often led to the same numerical results. But besides simplifying the mathematics for general consumption, their textbooks tended also to deny the role of judgment in any inferential method. This personal element was recognized by all statisticians. For Bayesians, it was declared in the choice of prior probabilities; for Fisherians in the construction of statistical model; for the Neyman–Pearson school in the selection of competing hypotheses. The social science texts, however, portrayed statistics as a purely impersonal and objective method for the design of experiments and the representation of knowledge.

---

[83] *Scientific Inference* of 1931 had no discernible influence on the social sciences. In order to turn the Bayesian method into a practical tool, Jeffreys proceeded to reproduce Fisher's estimation methods and significance tests from the epistemic standpoint. Even so, his magisterial *Theory of Probability*, published in 1939, suffered the same fate as *Scientific Inference*. Thus apart from hard-liners unwilling to see their discipline tainted by any suggestion of indeterminacy, the few social scientists who opposed Fisherian methods were not inverse probabilists but those who distrusted quantification more generally.

[84] Fisher wrote that "nearly all my statistical work is based on biological material and much of it has been undertaken merely to clear up difficulties in experimental technique." Fisher to A. Vassal, March 1929, as quoted in Box 1978, p. 17.

[85] Gigerenzer 1987, p. 18.

Gigerenzer has argued this was deliberate, necessary to produce what he called the "illusion of a mechanized inference process." In any event, the textbooks' apparently impersonal and universal 'statistics' included not just a number of Fisher's practical tools, such as his significance tests and regression analysis, but admixtures of Neyman–Pearson, such as the interpretation in terms of confidence intervals. Gigerenzer calls the result the 'hybrid' theory: "Statistics is treated as abstract truth, the monolithic logic of inductive inference."[86]

Strictly, the hybrid theory was based on incoherent foundations, and when reviewing these books statisticians grumbled. But not too loudly. The hope that statistics would be seen as a 'science of science' that could be universally employed as a general method of good thinking united statisticians from Pearson to Fisher.[87] Far better that social scientists slightly confuse the two approaches than not use statistics at all. Thus though occasionally sniffy, statisticians welcomed the statistical turn of the social sciences as enhancing the influence and thus professional autonomy of their own discipline. It would of course also improve the chances of individual statisticians finding employment as consultants or instructors.

6.3.3. The social sciences reformed for statistics

To become standard, a scientific instrument must be freed from its local experimental context. This depends not just on the usefulness and power of the instrument, but the persuasive and rhetorical strategies of its supporters. Gooday has shown that Victorian perceptions of the ammeter and voltmeter depended on the local status of the electromagnetic theory: practical electricians, comfortable with electromagnetic qualities and eager for robust and convenient measuring tools, were quick to adopt them; physicists thought they would mislead students about the real physical processes at work, while additionally inculcating bad research habits.[88] To reach consensus often

---

[86] See Gigerenzer 1989, pp. 106–9 (quote from p. 107); also Gigerenzer 1987, pp. 16–22. Gigerenzer notes that presentations of the hybrid theory routinely omitted the names of the responsible statisticians, and argues that this anonymity was a deliberate strategy to reinforce the objectivity and universality of the method. After the war, a few texts, such as the 1952 *Statistical Theory in Research* by Anderson and Bancroft, distinguished the Fisherian from the Neyman–Pearson approaches, but they remained in the minority: Neyman and Pearson were not mentioned at all in 21 of 25 texts Gigerenzer examined.

[87] Pearson, of course, saw statistics at the heart of scientific socialism too, a sentiment also expressed by H. G. Wells's comment that "Statistical thinking will one day be as necessary for efficient citizenship as the ability to read and write."

[88] Gooday 1995.

requires a transformation of the receiving discipline as well as of the device in question. For physicists to regard the ammeter and voltmeter as precision measuring instruments, not merely indicative tools, required both the discovery of the electron to raise electromagnetics to a 'fundamental' level, and a burgeoning of research to necessitate a production-line approach to teaching.

The same is to an extent true for less tangible products of science, such as numerical results and theoretical models. As we have seen in previous chapters, Fisher's statistics was more than a tool kit for analysis. It was the model of a particular sort of experimental science, and its standardization imposed on the social sciences a distinct style of experiment and interpretation. Fisher's recipe for an ideal scientific experiment was presented in *The Design of Experiments*. Though developed primarily for agricultural experiments at Rothamsted, the criteria of replication, randomization, blocking, and experimenter blindness soon came to characterize research in the social sciences. Analysis likewise became a matter of null-hypothesis testing. Research programs were thus transformed to fit the statistical model of the lottery mechanism: when properly gathered under specified conditions, data would form a standard set that could be imagined as a random sample from a fixed frequency distribution. The Fisherian approach became institutionalized, and alternative methods deemed invalid.[89] This was exacerbated by incomplete understanding of the theory. Fisher and Neyman repeatedly pointed out that the acceptance or rejection of the null hypothesis was never decisive, and that significance levels did not directly reflect the size of the effect under investigation. Yet social scientists habitually regarded an effect to be proved by the rejection of the null hypothesis. It became standard to interpret non-significant results literally, with the consequence that only 'significant' results tended to be published.[90]

---

[89] Gigerenzer cites as examples of such invalid alternatives Wundtian single-case experiments and the demonstrational experiments of the gestalt psychologists. He argues that by the early 1950s, the hybrid model was completely institutionalized in American experimental psychology. The focus on statistical technique also strengthened social scientists' behavioristic and instrumental assumptions. Porter too has considered the impact of statistics on the social sciences. His chapter on the subject in Grattan-Guinness 1994 concludes, p. 1340, that "Fisherian methods have come almost to define what constitutes sound research in the quantitative human sciences."

[90] As evidence of both the ubiquity of the Fisherian model of significance tests and its institutionalized misinterpretation, Gigerenzer 1987, pp. 21–22 points to A.W. Melton's editorial in a 1962 volume of the *Journal and Experimental Psychology*. The editorial outlines the journal's acceptance criteria in terms of null-hypothesis testing: an acceptance of the null hypothesis would almost guarantee the rejection of the paper for publication,

6.4. PROBABILITY IN PHYSICS

6.4.1. General remarks

Social scientists of the 1920s believed that statistics would give their studies the status of a full natural science. Ironically, the positivistic model of natural science to which they aspired was at the same time being challenged from within the hard sciences. The development of quantum mechanics marked an end both to physicists' longstanding commitment to mechanical causality, and to their claim to provide knowledge of a system independent of the experimenter. In this section I consider the effect of the quantum revolution on physicists' interpretations of probability. I will argue that, as in the social sciences, the frequency interpretation was reinforced. First, though, a rapid and somewhat tendentious sketch of the role of statistics and probability in the physical sciences up to around 1930.

6.4.2. Probability and determinism: statistical physics

As described in Chapter 2, it was primarily the social scientists of the nineteenth century who developed the study of statistics in order to tease causal patterns from the noise of their data. Because physical phenomena can usually be controlled to eliminate natural variation, the new methods found few applications in the natural sciences, apart from in error theory. Around 1850, however, chemists and physicists such as Rudolf Clausius and August Krönig started to relate the thermal properties of gases to the movements of their constituent molecules. Just as social scientists addressed aggregates of broadly similar but independently-acting people, the vast numbers of molecules that made up a volume of gas could likewise be considered 'statistical objects.' Though Clausius and Krönig typically assumed each molecule to move with the same speed, James Clerk Maxwell realized that collisions between molecules made a fully statistical approach necessary. Maxwell had been alerted to Quetelet's work by a review essay written in 1850 by the astronomer John Herschel, and believed that the fixed ratios found in social data justified the assumption that a sufficiently large number of independent molecules would exhibit a stable distribution of velocities. He derived this distribution in his first publication on the subject in 1860. It turned out to be the familiar normal curve, and he

since "the sensitivity of the experiment...was therefore not sufficient to persuade an expert in the area that the variable in question did not have an effect as great as other variables of known significant effect." A publishable paper was required to show rejection at the 0.01 level. See also Sterling 1959.

used it to make the surprising prediction, soon confirmed, that the viscosity of a gas was independent of density. In the following decade, Maxwell went on to produce theoretical expressions for transport phenomena such as diffusion rates, and for the mean free paths of molecules. Around 1870, Ludwig Boltzmann started work on the kinetic theory. Though like Maxwell using the results of social statistics to justify assuming a definite distribution of molecular velocities, he addressed himself to non-equilibrium conditions. In particular, he sought to explain the second law of thermodynamics, articulated by Rudolf Clausius in 1850: that heat always flows between real bodies to equalize their temperature. With his H-theorem of 1872, Boltzmann showed that the value of a certain function of the molecular velocities would always decrease until it reached a minimum given by the Maxwell distribution.[91]

What was the significance of this work for probability? Around 1870, most natural scientists still held that the unpredictability of nature was an illusion, and the concept of probability useful only because our knowledge of the world was incomplete. Indeed, the Laplacean form of determinism,[92] exemplified by Newton's account of the absolute regularity of planetary motion, had been fortified in the 1860s by Maxwell's unification of optical and electromagnetic effects into a simple set of differential equations. Statistical methods, when applied to social data, served to extend this culture of determinism to the human sphere: writers such as Buckle popularized the idea that observed regularities in, say, the suicide rate implied the existence of human laws that gave the lie to the impression of freedom. But the introduction of statistical methods to the physical sciences had a corrosive effect on the Laplacean doctrine, leading to what Hacking has called 'the erosion of determinism.'[93]

To start, the kinetic theorists justified their probabilistic assumptions as a practical convenience: individual molecules obeyed Newton's laws, but were so numerous in even the smallest volumes that a bottom up calculation of observed 'macrostates' of a gas from the motions of its molecules was impossible. The determinist Boltzmann had embarked on his studies of statistical physics to reconcile the irreversibility of the second law with his mechanistic philosophy. Thus though introducing probability directly into his description of non-equilibrium processes such as diffusion, he regarded it not as a reflection of chance, but as a mathematical trick that allowed the description

---

[91] See von Plato 1994, pp. 10–3, 77–90.

[92] Laplace wrote that "[a]ll events, even those which on account of their insignificance do not seem to follow the great laws of nature, are a result of it just as necessary as revolutions of the sun."

[93] This brief discussion of determinism is limited to the physical sciences; in the life sciences the debate revolved around markedly different axes.

of a large number of independent molecules. The change between molecular microstates depended on their probabilities, gases evolving to more probable states in a way completely determined by the H-function.

A number of scientists began to argue, however, that the concept of probability was necessary for a true description of nature, not merely a convenient shorthand for complexity. Maxwell, the first to describe explicitly a physical process in terms of probability, argued from the kinetic theory that knowledge of the physical world was always incomplete because it was statistical in character. It was impossible to determine the future state of a collection of molecules not only because their initial positions and momenta could never be known with sufficient accuracy, but because 'points of singularity' in the governing equations magnified even infinitesimal differences between these conditions into wholly different states. At such points it was a non-mechanical cause that decided the direction of a system's evolution.[94] So the introduction of statistics, which in the social sciences and psychology had fortified the belief in deterministic laws of human behavior, produced the opposite effect in the physical sciences. Science could no longer aspire to the certainty of Newtonianism; a degraded form of statistical knowledge might instead be the best available.

Maxwell's views on causality and determinism were colored by his strong religious beliefs. In a letter of 1867, he introduced the idea of an imaginary being who sat at the partition between two volumes of gas, opening and closing a small door to allow the faster molecules to pass into one side and the slower into the other. The actions of this 'demon' would result in one of the volumes of gas heating up while the other cooled, thus demonstrating that the second law was not a matter of certainty, but held only with a degree of probability. Maxwell argued that since the laws of thermodynamics were at some level a matter of chance, so Man's knowledge was necessarily imperfect. Natural science thus had definite limits which could be surpassed only with religious faith. Mechanical causes could not provide a complete account of nature; they were supplemented by hidden nonmechanical causes, thereby allowing for the possibility of free will without violation of natural law.[95] Yet mechanistically-minded scientists too believed in the necessity of probabilistic explanations. Boltzmann conceded as much in 1871. Poincaré went further. He espoused the epistemic interpretation of probability, speaking of an omniscient Mind that could perfectly foresee future events, and writing that "what is chance for

---

[94] See Porter 1986, pp. 205–6.
[95] Maxwell had been offended by Buckle's rejection of free will, because of its disturbing implications for moral responsibility.

the ignorant is not chance for the scientist. Chance is only the measure of our ignorance." But Poincaré regarded mankind's ever attaining such foresight as a theoretical as well as a practical impossibility. He pointed out that the mutual interactions of even three bodies could not be solved exactly, and like Maxwell argued that instabilities in the equations of dynamics and electromagnetics served to amplify small variations in initial conditions.[96]

6.4.3. Probability at the turn of the century

Though embracing probability as fundamental, neither Maxwell nor Poincaré believed that there was anything inherently random about the universe. Physical causes determined all events. Yet by the turn of the century it was increasingly difficult to explain away the appearance of chance. The second law of thermodynamics was continuing to resist deterministic explanation. As early as the mid 1870s, Thomson, and following him Johann Loschmidt, had brought up the problem of irreversibility: Newton's equations were symmetrical with respect to time, therefore could not by themselves account for the irreversible nature of observed thermodynamics. This objection was strengthened in 1896 by Ernst Zermelo, who used a theorem of his teacher, Poincaré, to prove that any purely mechanical system must always return to any previous state. Clearly, the universe could not be a purely mechanical system.

Boltzmann was forced to reply that the second law – and his H-theorem – was true only probabilistically that is, with near, but not complete, certainty. He conceded that in the fullness of time H must change symmetrically, meaning that any system would indeed return to its previous state, but argued that the initial conditions of the universe dictated that we live in a period of increasing disorder or entropy, and thus had an infinitesimal probability ever to observe such a change. But it was not just thermodynamic effects that were tainted by chance. Boltzmann had noted in 1896 that even when in equilibrium, a statistical system was necessarily subject to chance fluctuations. Though he expected these never to be exhibited in a gas, since they would be averaged out across the vast numbers of molecules in a macroscopic volume, random fluctuations had already been suggested in the 1880s as an explanation of Brownian motion. The phenomenon – the perpetual and chaotic motion of small particles in suspension – had been discovered in a sample of pollen

---

[96] See von Plato 1994, pp. 42–3; 170–1. Emile Borel followed Poincaré in basing an argument that determinism was meaningless as a scientific concept on the instabilities inherent to mechanical systems.

in 1827 by the botanist Robert Brown, and had resisted several attempts to explain it, for instance as an effect of electrostatic charge or the dissolving process. A remaining explanatory candidate was that the motion was due to fluctuations in the numbers of molecules buffeting each side of the particle. Working from this assumption, Albert Einstein gave a complete statistical account of the effect in 1905, and his model was confirmed by experiments carried out by Jean Perrin around 1909.[97]

The phenomenon of radioactivity, resulting from the spontaneous disintegration of atomic nuclei, was discovered by Becquerel in 1896, and also appeared to involve chance. Nuclei seemed to decay at random, rather than under the influence of, say, electromagnetic fields or conditions of temperature and pressure. In 1900, Rutherford showed that the rate of decay was proportional to the size of a sample, consistent with each atom having a fixed chance of decay per time interval. With Soddy in 1903 he tentatively proposed the effect to be random rather than caused, and seven years later with Geiger showed that this assumption could also explain the observed fluctuations around half-life decay curves.

Most physicists, however, though prepared to accept probabilities for physical description, continued to cling to the idea that the universe was completely lawlike. They rejected Rutherford and Soddy's interpretation of radioactivity, and notwithstanding the atomic model – electrons orbiting a stable nucleus like planets round the sun – tried to explain the phenomenon as a consequence of instabilities in the atomic system.[98] Probability was also fundamental but not irreducible in early versions of the quantum theory. In 1900, Max Planck published his derivation of the law for black body radiation, the distribution across wavelengths of the energy radiated from an idealized black surface. Borrowing Boltzmann's assumption of discrete, or quantized, energy levels, he used combinatorial analysis to obtain a distribution that was similar to the form empirically observed in approximate black bodies. Planck regarded this use of probability, however, as just a calculational convenience. Einstein's position was similar. He was quick to adopt the quantum theory, accounting for the photoelectric effect in 1905 by treating a radiation field as discrete photons in a cavity, like molecules in a container of gas. Probability was central

---

[97] Einstein's explanation has been taken as final proof that matter is made up of discrete molecules and atoms.

[98] Since atomic nuclei do not interact with each other, the 'chance effect' had to come from within the atom or nucleus, rather than emerge as a consequence of the collective, as in kinetic theory. F.A. Lindemann in 1915 presented decay as a consequence of calculable – in principle, at least – coincidences in the independent movements of sub-particles within the nucleus. See Gigerenzer 1989, pp. 192–3.

to much of his work. In addition to the theory of Brownian motion mentioned earlier, Einstein used a probabilistic treatment of atomic transitions in 1916 to recover Planck's law from Bohr's atomic model.[99] Yet Einstein's philosophy broadly followed that of Poincaré: probability was an epistemic feature, not part of the furniture of the world, and its presence in his theory was a weakness that indicated an incomplete understanding of the causal mechanisms at work. Einstein regarded his 'spontaneous' transition probabilities as provisional, hopefully to be eliminated in a more fundamental description of nature and replaced by a determinate factor. (In a paper of 1917 on the subject he left the word 'chance' in quotation marks: "Zufall.") His work in relativity can be regarded as similarly conservative, a reconciliation of Maxwellian electromagnetism with classical, deterministic mechanics.

It was also unclear precisely what such probabilities meant. They had been introduced in kinetic theory to counter the difficulty of describing a system made up of molecules, and with the blessing of social statistics were equated to stable ratios. Yet these ratios admitted of various interpretations. Boltzmann took the frequentist view: the probability of a velocity was the average time a particular molecule spent traveling at that velocity. But a more epistemic interpretation was also possible. In the early 1870s, Maxwell introduced the idea of the ensemble, an imaginary collection of systems each with a different mechanical microstate but of the same total energy. Here a probability is the ratio of possible microstates consistent with the observed macrostate, and reflects imperfect knowledge of which particular microstate the system is in. Most kinetic theorists, however, were more concerned to make progress with the statistical model than to agonize over the meaning of probability. Josiah Gibbs is now firmly associated with the ensemble approach, having developed Maxwell's idea in his 1902 study, *Elementary Principles in Statistical Mechanics*. Yet at the time he moved freely between the epistemic, objective-frequency, and even classical interpretations. Einstein too, though regarding probability as evidence of incomplete description, was content to quantify from empirical time-averages.

6.4.4. The rejection of causality

Despite Einstein's efforts, the old quantum theory proved stubbornly incompatible with the classical model of the atom. From around 1920, physicists such as Franz Exner and Hermann Weyl began to reject Newtonian

---

[99] Bohr produced his quantized atomic model of 1913 to account for the discrete rather than continuous nature of atomic transition spectra.

mechanism, and by the end of 1926 a fully indeterministic theory was available in the new quantum mechanics. The full story of the years 1925 and 1926 is complex.[100] Though probability was fundamental to both Heisenberg's 1925 version of the theory and the subsequent matrix form of Max Born and Pascual Jordan, most physicists initially preferred the more tractable differential equations of Schrödinger's wave account of early 1926. The continuity suggested by these equations was comfortingly familiar, despite the unquestionably discontinuous nature of quantum processes such as atomic transitions. But how to interpret the wave function? Schrödinger proposed that the square of the modulus be equated with charge density, but could not explain how dispersive waves could adequately represent discrete particles. Heisenberg criticized these attempts as reactionary, and used the positivist language of the Vienna School to call for such classical notions as electron orbits and trajectories – and the concomitant problem of atomic stability – to be abandoned. Instead, the probability numbers of the matrix formulation should be regarded as the only meaningful properties. A compromise was reached by Born. He interpreted the wave function directly as a sort of probability, and showed in mid 1926 that this was the only conclusion consistent with Schrödinger's wave account. Probability, he declared publicly, was fundamental, and "[f]rom the point of view of quantum mechanics there exists no quantity which in an individual case causally determines the effect of a collision."[101]

Schrödinger soon demonstrated his and Heisenberg's theories to be equivalent, but remained uneasy at Born's probabilistic explication of the wave account. In October of 1926 he traveled to Copenhagen, where he met with Bohr, Heisenberg, and the British theorist Paul Dirac. Bohr had renounced classical descriptions of atomic transitions soon after introducing his quantized model of the atom in 1913, and persuaded the other physicists toward Born's view. He argued that observables such as position and momentum were not fundamental, but emergent properties of the irreducible probability distribution described by the Schrödinger equation. These probabilities could be measured only as statistical frequencies, meaning that individual events were not causally related. Heisenberg's 1927 paper on the uncertainty principle – ill-named: 'indeterminacy principle' would have been better – cemented the position. Not only was chance part of the stuff of the universe, but according to a notion Bohr called 'complementarity,' the old distinction between the experimenter and the system studied was arbitrary, the two being

[100] See, e.g., the multi-volume work Mehra and Rechenberg 1982–7.
[101] Born as quoted in von Plato 1994, p. 152. See von Plato 1994, pp. 93–114 for the relationship between ergodic theory and quantum mechanics.

entangled in a quantum state. Mechanistic causality was thus an illusion; the world was inherently probabilistic. This interpretation, officially at least, was quickly accepted.[102]

6.4.5. The view in the 1930s

Historians have studied the reception of quantum mechanics almost as much as its formal development. The radical and counter-intuitive nature of the theory certainly made it a good candidate for selective interpretation, according to an individual's expectations, prejudices, and intellectual environment. Paul Forman has famously argued that their defeat in World War I turned Germans from a reverence for technology and rational thinking to a pessimistic sort of "celebration of 'life,' intuition, unmediated and unanalyzed experience," as exemplified by the historicism of Oswald Spengler and the intuitionism of the Dutchman L.E.J. Brouwer. Physicists, who had quickly adapted to this atmosphere, proposing severally from around 1919 that the formal and analytical foundations of physical science be reformulated along non-causal lines, eagerly seized the probabilistic interpretation of quantum mechanics as justification for the rejection of abstract rationality as barren and culture-bound.[103]

Forman's argument that theories are interpreted according to the local context explains why the 'missionaries of the Copenhagen spirit' had a job carrying their views elsewhere.[104] While the continental version of quantum mechanics was strongly marked by metaphysical language – Bohr's thumb prints being particularly evident – American scientists were largely indifferent to such implications. Nancy Cartwright has documented this lack of concern, noting that in America, "indeterminism becomes codified in quantum mechanics almost without remark."[105] Attendees at a symposium on quantum mechanics for the American Philosophical Society in 1928

---

[102] Supporters of the new quantum mechanics pointed to historical precedents for the probabilistic view. Bohr argued that the kinetic theory had made the eventual acceptance of probabilism inevitable; Jordan that the case for indeterminacy had been proved by the evidence of radioactivity. But van Brakel 1985 has made a convincing historical case that radioactivity was only seriously adduced as an argument for indeterminacy after 1928.

[103] Forman 1971; see also Forman 1979 (and Hendry 1980 for a critical response). The quotation is from Forman 1979, p. 13. Forman gives as examples of those who rejected causality, in order, Franz Exner, Hermann Weyl, Richard von Mises, Walter Schottky, Walter Nernst, Erwin Schrödinger, Arnold Sommerfeld, Hans Albrecht Senftleben, and Hans Reichenbach.

[104] See, for example, Heilbron 1988.

[105] Cartwright 1987.

were thus untroubled by the role of probability; young physicists who used the theory, such as Oppenheimer, Condon, van Vleck, Slater, and Rabi, regarded such metaphysical anxieties as a European preserve. Heisenberg noted in 1929 that "most American physicists seemed prepared to accept the novel approach without too many reservations."[106] It was not that American physicists were uninterested in foundational issues, Cartwright argues, rather that they were insulated by their culture of pragmatism – especially the operationalist version Bridgman popularized among physicists – against woollier issues such as the conflict between free will and determinism.

Elsewhere things were different again. Unlike in Germany, the prestige of science in England had been heightened by the war. Newspapers and magazines were stiff with popular accounts of atomic theory, stars and the universe, and relativity. Forman has studied the reception of quantum theory in Britain up to 1927. He finds that their confidence in the scientific world view gave British scientists a strong cultural commitment to determinism. The few hints of anti-mechanism among scientists and philosophers prior to 1925 were not marked by a strong acausal propensity; indeed, they were an expression not of the retreat of a cowed physical science, but its triumphant advance into the realms of psychology and spirituality.[107] Consequently, British scientists tended to miss the acausal implications of quantum mechanics. This was true even of innovative thinkers. Norman Campbell – one of Jeffreys's partners in the establishment of the National Union of Scientific Workers – had rejected the logical formalism of philosophers' attempts to account for scientific knowledge, in favor of an account based on the actions and instincts of experimental scientists.[108] He declared that time was a statistical average across atomic transition probabilities, related in a similar way as temperature to molecular velocities. As Forman notes, however, he shied from linking these views explicitly to acausality or indeterminism.[109]

Arthur Eddington was a notable exception. First Wrangler at Trinity in 1904, and going on to produce ground-breaking theoretical work on the structure and evolution of stars,[110] he was also a practical astronomer who had cataloged numerous transit observations and led experimental expeditions, including that of 1919 to photograph the solar eclipse. But his view of science had a fanciful side. In 1925, he proposed a connection between physics and consciousness: determinism was a mental artifact imposed on

---

[106] Quoted in Cartwright 1987, p. 418.
[107] Forman 1979, pp. 26–30.
[108] Campbell 1919.
[109] Forman 1979, pp. 31–2, 37.
[110] Eddington pioneered the view that stellar energy comes from the conversion of matter.

the world; freedom from it would reveal a deep connection between the elementary particles and the animating spirit of the mind. Eddington was quick to embrace the acausal implications of quantum mechanics. Lecturing on the relation between science and religion in 1927, he announced that the new theory removed determinism from the laws of physics, and that "whatever view we may take of free will on philosophical grounds, we cannot appeal to physics against it."[111] But Eddington remained a solitary figure, unsurprising in the light of the still wilder turn his views took during the 1930s. Fascinated by the fundamental constants, he believed that each could be derived *a priori* from a process of logic and introspection, and linked with the theories of relativity and quantum mechanics. Again, this was licensed in part by the extension of the mind's domain indicated by the uncertainty principle.

Eddington's 'selective subjectivism' held that it was the power of our minds and sensory equipment that decomposed the universe into types and numbers of particles. "I believe that the whole system of fundamental hypotheses can be replaced by epistemological principles. Or, to put it equivalently, all the laws of nature that are usually classed as fundamental can be seen wholly from epistemological considerations. They correspond to *a priori* knowledge, and are therefore *wholly subjective*."[112]

So Eddington became a prominent but isolated English advocate of indeterminism through the 1930s.[113] His position, like that of the German-speaking quantum physicists, was "an expression of religious-philosophical yearnings."[114] Unlike the Germans, however, Eddington's indeterminism seems to have grown more from his studies of relativity – of which he was an early defender – than from quantum mechanics. In any case, the gradual rejection of determinism by British physicists through the 1930s seems unconnected with mysticism of the Eddington or German varieties. And more than one prominent scientist held out against the full probabilistic interpretation. The astrophysicist Herbert Dingle, for example, reacting against Eddington's esoterica, reluctantly conceded a position for probability, but denied that it signified a wholesale rejection of determinism. "So far from involving caprice,

---

[111] Quoted in Forman 1979, p. 38.
[112] Eddington 1939, pp. 56–7; see also p. 60. Eddington's mysticism quickly made him a figure of fun to many physicists. In a spoof paper, three German physicists claimed to have derived the value of 137 for the reciprocal of the fine structure constant from general considerations of vibrations of a crystal lattice. This was the number of degrees of freedom Eddington claimed for the electron. The paper slipped past the editors and appeared in a 1931 issue of the journal *Die Naturwissenshaften*. See Weber 1973, p. 24.
[113] See Eddington 1935.
[114] Forman 1979, p. 34.

it symbolizes the factor in the atom which excludes caprice . . . No amount of theorizing can alter the observed fact that there is a determinism in nature."[115] Experiment revealed that the atomic world behaved very differently from the macroscopic one, and that mechanical notions such as position and momentum do not apply cleanly to fundamental particles. In order to persevere with these irrelevant qualities, we are forced to introduce the concept of probability. But this should not fool us into thinking that we have discovered something about human experience. Probability is merely a form of expression. "Words like *uncertainty* and *probability*, ideas like the necessity of interference of subject with object, are presented as indications of some profound characteristic of nature instead of as tentative efforts to express correlations which are as definite and certain as any others in science."[116]

6.4.6. The interpretation of probability in physics

Despite this range of attitudes to quantum mechanics, the rise of the theory during the 1930s accustomed physicists to regard probabilities as fundamental. Since these probabilities seemed to be a feature of nature, they tended to be interpreted statistically. An indeterminist sees a world of chance, naturally to be quantified with frequencies. Thus Paul Dirac wrote in 1930 that when "an observation is made on any atomic system . . . the result will not in general be determinate, i.e., if the experiment is repeated several times under identical conditions several results may be obtained. If the experiment is repeated a large number of times it will be found that each particular result will be obtained a definite fraction of the total number of times, so that one can say there is a definite probability of its being obtained any time the experiment is performed."[117] Yet even those dubious about indeterminism recognized that quantum mechanical probabilities were more objective and foundational than the old degrees-of-belief. For Dingle a probability was a frequency, to be contrasted with "the vague quality which depends on individual predisposition."[118] Campbell too was slow to relinquish determinism. Yet he also regarded the epistemic interpretation as irrelevant to his model of science (which he restricted to those areas of inquiry on which universal agreement obtained). Probability did have a scientific role, but only as an objective model

---

[115] Dingle 1931, p. 87. Schrödinger retained a similar belief in determinism, with the wave function rather than classical descriptive categories as the underlying reality.
[116] Dingle 1937, p. 301.
[117] Dirac 1930.
[118] Dingle 1931, p. 46.

of nature, to be determined by experiment from a large number of trials.[119] Von Neumann was plain in his 1932 work on quantum mechanics. Discussing the "problem as to whether it is possible to trace the statistical character of quantum mechanics to an ambiguity (i.e., incompleteness) in our description of nature," he noted that "such an interpretation would be a natural concomitant of the general principle that each probability statement arises from the incompleteness of our knowledge," and that this "explanation by 'hidden parameters'. . . has been proposed more than once." But he concluded that "such an explanation is incompatible with certain qualitative fundamental postulates of quantum mechanics," which therefore was "in compelling logical contradiction with causality." Probability is fundamental, and statistical ensembles are necessary for "establishing probability theory as the theory of frequency."[120] Von Neumann's operator theory of quantum mechanics was widely adopted, and his 'no hidden variable' theorem taken by the mathematically-inclined to rule out the possibility of any as-yet undiscovered causation.

Perhaps, though, Bohr's Copenhagen interpretation of quantum mechanics, with its emphasis on the entanglement between observer and observed, could support the link between probability and knowledge? Probability might be irreducible, the stuff of nature, but with the line between consciousness and matter blurred, it could also be a quality of the mind. After all, Heisenberg occasionally spoke of the uncertainty principle in terms of experimental observations 'destroying the knowledge' of a system.[121] In practice, however, even the most extreme panpsychics did not call for quantum mechanical

---

[119] Though Jeffreys approved of much of Campbell's exposition, he vigorously opposed the criterion of universal assent. Wrinch and Jeffreys 1921a, p. 374: "Science therefore rests on individual judgments, and so far is similar to mysticism or art; the judgments that electric potential satisfies Laplace's equation and that 'The Magic Flute' is a superior work to 'The Bohemian Girl' are distinguished by our own feelings about them and not by any external criterion. Even the *a priori* postulates, in their actual application, are individual judgments." Jeffreys 1923b, p. 1021 argued that Campbell's criterion missed the essence of science, since it is "just precisely the propositions that are in doubt that constitute the most interesting part of science; every scientific advance involves a transition from complete ignorance, through a stage of partial knowledge based on evidence becoming gradually more conclusive, to the stage of practical certainty." Jeffreys also disagreed with Campbell's view of the status of causal laws: "I do not believe that [scientific laws] deal with certainty, for if they did there should be a stage in one's knowledge about a law when it passes suddenly from complete ignorance to complete certainty, which does not appear to be the case . . . " But Campbell was not prepared to accept Jeffreys's definition: "You say that my definition of probability is scientifically inapplicable; if by that you mean inapplicable to the probability of propositions, of course it is; that is why I chose that definition." Jeffreys 1923b, p. 1023; Campbell to Jeffreys, 21 October 1921; HJ.
[120] von Neumann 1955 [1932], pp. x, 298, 327; see also pp. 313–327.
[121] See, e.g., von Plato 1994, p. 159.

probabilities to be glossed as epistemic. Eddington believed that a probabilistic approach was essential both to describe our interaction with the universe and as a reflection of knowledge, but took such knowledge as an objective feature of the indeterministic system of Mind and Universe, not a measure of a particular individual's degree of belief. Quantum mechanical probabilities were statistical, to be derived from observation: "Whatever significance probability may have in other departments of thought, in physical science probability is essentially a statistical conception . . ."[122] Eddington did not in fact believe inference to be a numerical process. He rejected inverse probability as "silly," and the Principle of Insufficient Reason in particular as "the height of absurdity."[123]

6.4.7. Quantum mechanics and inverse probability

The rise of quantum mechanics, and the associated rejection of determinism, goes some way to explaining why inverse probability fell from favor during the 1930s. A Bayesian does not rule out stochastic chance in the world, but in concentrating on probability as a representation of imperfect knowledge, tends to chalk up uncertainty to the investigation process. He is obliged to regard the result of a single coin toss as a matter of significance rather than luck, for example, and to amend his original prior probability assessment accordingly.[124] The position thus has a natural sympathy with determinism. Certainly this is not a matter of necessity or logical entailment. But it does seem to square with the views of many adherents of the epistemic position in the 1920s and 1930s. Keynes, for example, adopted the Laplacean view, regarding a 'chance' event as one "brought about by the coincidence of forces and circumstances so numerous and complex that knowledge sufficient for its prediction is of a kind altogether out of our reach."[125] He thought it 'irrational' for frequentists simply to assume such events indeterminate; we must assess them with our intuition and good judgment. Jeffreys took a similar view. Even if some events *were* random, their existence could only be inferred through considerations of probability. This is why the errors he discussed in *Scientific Inference* were not random, but due in each case to determined if unknown factors.

---

[122] Eddington 1939, p. 95.
[123] Eddington 1933, p. 274; Eddington 1935, p. 112.
[124] Thus for a naive Bayesian, the hypothesis of bias will be reinforced after a single toss whatever its result.
[125] Keynes [1973], p. 326.

Quantum mechanics hence presented problems for the inverse probabilist.[126] Jeffreys was skeptical of the quantum theorists' claims for acausality, denying in the 1937 reissue of *Scientific Inference* that Heisenberg's uncertainty principle had killed determinism in the sense normally meant.[127] After all, that the most refined experiments known to modern physics were limited in accuracy already followed from the ever-present variations in our sense-experience. Jeffreys expanded on the point a few years later. A particle's position and momentum are never known exactly because of experimental error; thus to predict future states in a classical system, a joint probability over the coordinates is required. In the case of the Darwin solution of a free electron in a uniform field, "the whole uncertainty of the prediction is traceable to that of the initial conditions [of the electron]... and there is nothing in the solution to indicate that this is less deterministic than in classical mechanics."[128] Not only is the quantum theorists' introduction of probability hardly novel, he argued, but their talk of 'underlying indeterminacy' is unverifiable and therefore meaningless. Probabilities are limits on our knowledge rather than intrinsic to the atom: "To try and represent [the momentum] by a single continuous probability distribution is impossible; given an observation there is one distribution for the probability before the observation and another after it, and the two cannot be combined, being on different data... The demand in wave mechanics that description be restricted to observed quantities is impossible to satisfy... The steady state is not observed as such but inferred; I contend that it is a legitimate inference, and that it can be described in terms of exact orbits."[129] In fact, Jeffreys doubted whether quantum theory relied on probability at all, and bridled at descriptions of Schrödinger's $|\psi|^2$ as a probability density. Scientific laws relate observables, yet these quantum mechanical 'probabilities' involve complex numbers and non-commutative multiplication; also "it is certainly meaningless to speak of a probability without specifying the data, and although functions obtained in quantum theory

---

[126] The inverse probabilist of the old school, at least. As discussed in the next chapter, de Finetti developed a theory of subjective probability in direct response to the indeterminism of quantum mechanics.

[127] Jeffreys [1937], pp. 246–48.

[128] Jeffreys 1942, p. 818. See also p. 824: "the quantum theory is deterministic in the classical sense; and indeterminacy arises from imperfect knowledge of the initial state, and not from the laws of motion."

[129] Jeffreys 1942, p. 829. Jeffreys conceded, however, that determinism must be 'relaxed' slightly, and a degree of randomness permitted, to account for radioactive decay and spontaneous transition, see pp. 829–30.

are interpreted as probabilities, there are hardly any cases where the data are properly specified."[130]

6.5. PROBABILITY IN BIOLOGY

Biologists continued to debate evolution through the 1930s. Fisher's views remained steadfast. He insisted – as he had when an undergraduate at Cambridge – that Darwin and Mendel's systems were not only compatible but complementary. Traits resulting from many genes display blending inheritance, as the biometricians had found. Yet an appeal to environmental effects to counter the blending of adaptive modifications is unnecessary, since Mendelian inheritance preserves variation. Natural selection, Fisher declared, was the sole mechanism of evolution.

Fisher developed his position in *The Genetical Theory of Natural Selection* of 1930. It is a rich book. In Chapter 2 he presents his "Fundamental Theory of Natural Selection," which relates the mean fitness of a species to its genetic variance; elsewhere he discusses environmental effects on genetic variation, and produces mathematical accounts of gene survival and spread, and mutation rates and effects. He explains the theory of dominance, developed from breeding experiments at Rothamsted, and his comments on sexual selection provide an evolutionary explanation for the preponderance of male over female births (the observation used by John Arbuthnot in 1710 for his probabilistic argument for divine providence).

The evidence for Mendelian evolution continued to accumulate through the decade.[131] Yet Fisher's views remained controversial. It is revealing that he barely touched on the contemporary work of T.H. Morgan's group on drosophila chromosomes, for example. In part, this was because he was not so interested in purely biological developments of genetic theory; in part

---

[130] Jeffreys 1942, p. 819; see also p. 828.
[131] Though Dobzhansky popularized the corn experiments of Floyd Winter as experimental evidence for genetical evolution in his book of 1937, Student was the first to draw attention to the work. From 1896 to 1924, Winter had selectively bred two strains of corn, one with a mean percentage of oil content twelve standard deviations (of the original stock) above the original stock mean; the other seven below. His results, published in 1929, were mentioned by 'Student' in a letter to Fisher of 26 January, 1932: "And so we reach the conception of a species patiently accumulating a store of genes, of no value under existing conditions and for the most part neutralised by other genes of the opposite sign. When, however, conditions change ... the species finds in this store genes which give rise to just the variation which will enable it to adapt itself to the change." Fisher persuaded 'Student' to write this up into a note, and had it published in *The Annals of Eugenics* after rejection by *The American Naturalist*. See E.S. Pearson 1990, pp. 66–7.

because as a statistician and indeterminist he was concerned more to investigate the consequences of gene frequencies for heredity than to explain these frequencies by mapping chromosomes.[132] But it was also because Fisher believed in the exclusive influence of natural selection. In contrast, Mendelians such as Punnett and Goldschmidt held that continuous variation was not heritable, and merely produced fluctuations around a main line of evolution due to mutation. Fisher replied that mutations were directionless and too infrequent to have a significant effect on evolution; Morgan's work indicated they were generally harmful in any case.

An analogous issue marked the dispute between Fisher and the American evolutionary theorist, Sewall Wright. In addition to mutation, which introduces random disorder at the level of the individual, the mechanism of 'genetic drift' introduces disorder at the level of the population. Proposed by Wright in 1931, genetic drift is a sampling effect.[133] Genes pass from one generation to the next by random selection. The frequency of genes in a large population will remain approximately the same between generations, but may differ if the population is small. As illustration, consider sampling with replacement from an urn of colored balls. After n draws, a second urn is filled with balls corresponding to the colors recorded. If n is large, the ratio of colors in the second urn will be almost identical to the first. But if n is small, the contents of the two urns might differ significantly.

Fisher disliked the idea that random processes could produce order without guidance from natural selection. He likened genetic drift to thermodynamic dissipation, and endeavored to show that like mutation it was negligible as an evolutionary force. Fisher tended to assume large populations, also necessary for his theory of dominance. Wright, on the other hand, in common with many field biologists whose work had exposed them to small sub-species, imagined isolated pockets of individuals, split from each other by geographical conditions. The gene pools of such groups would diverge in time due to genetic drift, and unless there was a large selective pressure on the traits in question, changing circumstances could result in selection at the level not of the individual but of the whole group.

Fisher differed with Wright over the importance of genetic drift in nature. But though their dispute was as vitriolic as Fisher's contemporaneous clash with Neyman, it was a dispute over the extent, not the role or interpretation

---

[132] Morgan and his group sought to explain and predict observed ratios of traits by mapping the locations of genes along chromosomes. They assumed chance association, though modified due to cross-over and linkage effects. See Kohler 1994.

[133] Wright 1931.

of probability. Both Fisher and Wright based their models on random selection from distributions of probabilities, and both recognized random drift
as unquestionably a part of this theory. The genetic drift debate amounts
to a difference over the value of a certain empirical parameter, and is still
unresolved.[134] But in the broad picture, the extension of the genetic model
reinforced the importance of probability. The 'evolutionary synthesis' hammered out during the 1930s – credited jointly to Fisher, Wright, and Haldane –
was like the new quantum mechanics a fully stochastic theory, and the chances
involved had clear interpretation as frequencies.

Fisherian statistical methods, especially significance tests, accordingly became institutionalized in biological research. Genetic examples padded successive editions of *Statistical Methods for Research Workers*. Fisher showed
clearly how his methods could be used in practice, for example to determine
whether differences between observations and expected ratios were probably
a result of natural variation, or whether the Mendelian hypothesis in question
should be modified. During the 1930s, he developed a number of statistical
techniques specifically addressed to population genetics. This focus in fact
strengthened the analogies to his model of experimental science. The infinite hypothetical population can be likened to the gene pool, and a particular
experimental result considered a chance selection from a fixed distribution
of probability in a similar way that an organism's genotype arises by chance
selection from the stable frequencies in the species' gene pool.

A final remark. I have argued that among physicists of the 1930s there was a
rough correlation between views on probability and attitudes to determinism,
frequentists tending to be indeterminists and Bayesians determinists. This is a
shaky association – further confused by Bohr's doctrine of complementarity –
and its extension to biology still more precarious. For experimental biologists, the magnitude of natural variation makes the choice between physical

---

[134] After World War II, while field biologists provided much evidence for the importance
of small, often isolated sub-species, 'selectionists' (i.e., those like Fisher who rejected
all but natural selection as a directional factor in evolution) demonstrated that many
of the variations between groups ascribed to random drift actually conferred adaptive
advantages, and dismissed the mechanism as an empty attempt to explain phenomena
when no obvious adaptive reason for variation could be found. During the 1960s, however,
molecular biologists showed that such non-adaptive variations could occur due to chance
mutation at the DNA level. As well as being a turf war between sub-disciplines, and
between theorists and practitioners, this debate can also be viewed nationally. Turner
1987, p. 314, notes that "relegation of random drift to the role of a trivial 'fluctuation'
in evolution has come to characterize the distinctive English national style in population
genetics." Mayr and Provine 1980 is a collection of essays that examine different attitudes
to the evolutionary synthesis between disciplines and countries.

determinism and indeterminism a practical irrelevance. Also, though the frequency interpretation is suggested by the notion of randomness that underlies the evolutionary synthesis, there is still room for epistemic probability, since the direction of evolution depends on environmental change due to definite but unknown causes. Yet both Fisher and Wright were indeterminists, albeit reluctantly in Wright's case, and both drew support for irreducible chance from results in the physical sciences, especially the kinetic theory.[135]

6.6. PROBABILITY IN MATHEMATICS

6.6.1. Richard von Mises's theory

Toward the end of the nineteenth century, the Russian school of Pafnutii Tchebychev and A.A. Markov decoupled probability theory from its social and astronomical applications, and began to address it as a branch of pure mathematics. The German David Hilbert furthered the process, including in 1900 the axiomatization of probability theory as the sixth of his famous list of twenty-three unsolved problems in pure mathematics. In the following decades, a rash of probability work appeared in the mathematics journals. Contrasting with the empirical tradition found in the U.K. and America, these attempts to put the theory on sound foundations were chiefly due to continental thinkers, such as Emile Borel.[136] One of the most important was the frequentist theory of the German Richard von Mises.

Lax 'proofs' of such as the central limit theorem provoked von Mises to review probability theory around 1918. Von Mises was not just a mathematician. He wrote a seminal text on hydrodynamics, was a pioneer of aviation, and was the model for the character Ulrich in Musil's modernist classic *Der Mann ohne Eigenschaften*. He was also a member of the Vienna Circle, and his logical positivism influenced his approach to probability and statistics. Science was the search for the most economical description of the universe, he insisted, which meant avoiding postulating superfluous causes. The kinetic theorists' attempts to justify their statistical theory in terms of classical mechanical concepts such as collision and trajectories was pointless.[137] It also conflicted with

---

[135] See Turner 1987; also Gigerenzer 1989, pp. 160–2.

[136] See von Plato 1994, Chapter 2.

[137] von Mises [1957], p. 223: "The point of view that statistical theories are merely temporary explanations, in contrast to the final deterministic ones which alone satisfy our desire for causality, is nothing but a prejudice. Such an opinion can be explained historically, but it is bound to disappear with increased understanding."

thermodynamics. Better to base a statistical theory directly on probabilities and pure mathematics.

All theories require some primitive postulates. Fisher assumed a stable limiting frequency and the infinite hypothetical population, and Jeffreys that probability numbers express degrees of belief. Von Mises, whose reading in thermodynamics had inclined him to indeterminism,[138] based his theory on the concept of randomness. He introduced the *kollektiv*, an infinite set of observations with the property that frequencies of each observed attribute within the set have definite limits. The collective is random, defined to mean that these limits remain the same for subsets of attributes formed according to any predetermined selection rule.[139] Probabilities were the limits; they existed only for the entire collective. Von Mises's theory of probability, as contained in his books *Wahrscheinlichkeit, Statistik und Wahrheit* of 1928 and *Wahrscheinlichkeitsrechnung* of 1931, consisted of rules for the combination of collectives.[140]

Von Mises's collectives were abstractions, his theory an idealization. Could they be implemented in practice? Only in repeatable experiments that could yield an infinite sequence of random results with fixed frequencies. Likely candidates included games of chance – for which von Mises regarded the assumption of limits as most reasonable – and certain matters of biological inheritance[141]; he also allowed cautious applications to some problems of mortality. Nevertheless, the theory was restrictive, perhaps befitting a product of pure mathematics. Unlike both Fisher and Jeffreys, who saw probability theory as a general scheme of rationality, von Mises regarded it as a science, not experimental but "of the same order as geometry or theoretical mathematics."[142] In consequence, he had no time for probability as an index of subjective belief. True, the word was colloquially used in this sense, but science is a matter of refining such vague and loose usage. Von Mises was

---

[138] von Mises [1957], p. 223: "The assumption that a statistical theory in macrophysics is compatible with a deterministic theory in microphysics is contrary to the conception of probability expressed in these lectures."

[139] Alonzo Church extended this idea of randomness in 1940. See Church 1940.

[140] *Wahrscheinlichkeit, Statistik und Wahrheit* was published by Springer as part of a series by members of the Vienna Circle. *Wahrscheinlichkeitsrechnung* was a more ambitious work, and it was this, rather than Jeffreys's *Scientific Inference* of the same year, that prompted Haldane's 1932 paper on sampling. Von Mises's idea of collectives was influenced by Fechner.

[141] von Mises [1957], p. 161: The success of Mendel's theory "is one of the most impressive illustrations of the usefulness of the calculus of probabilities founded on the concept of the collective."

[142] von Mises [1957], p. v.

fond of the analogy to thermometry, a study that began when impressions of temperature were replaced by an impersonal scale of comparison based on the expansion of mercury. "Everyone knows that objective temperature measurements do not always confirm our subjective feeling, since our subjective estimate is often affected by influences of a psychological or physiological character. These discrepancies certainly do not impair the usefulness of physical thermodynamics, and nobody thinks of altering thermodynamics in order to make it agree with subjective impressions of hot and cold."[143]

Von Mises treated with contempt any attempt to turn frequency into a measure of plausibility. Fisher's likelihood was an example. ("I do not understand the many beautiful words used by Fisher and his followers in support of the likelihood theory."[144]) He was particularly harsh about Fisher's work on small samples. Probabilistic inference could indeed be carried out with Bayes's theorem, but only from one collective to another.[145] Von Mises's probabilities referred to the entirety of a long sequence of experiments. Talking of the next event in a collective being 'almost certain' was "not too reprehensible so long as we realize that it is only an abbreviation and that its real meaning is that the event occurs almost always in a long sequence of observations."[146] Thus small samples were useless, the theory "to be completely rejected."[147]

Von Mises was respected as an applied mathematician, but his foundational work on probability attracted mostly negative comment. His collectives were not physically realizable, his limits an unjustified assumption, his definition of randomness precarious and inadequate, his theory overly narrow.[148] Such criticisms became increasingly vocal through the 1930s as the importance of mathematical probability for nuclear and quantum physics became clearer.[149] Yet the core frequentism of von Mises's view was unchallenged

[143] von Mises [1957], p. 76.
[144] von Mises [1957], p. 158.
[145] Prior probabilities were not a problem for von Mises, since he believed that Bayes's theorem applied only to infinite sets of data. (Posterior probabilities converge for large quantities of data whatever the prior distribution.) Von Mises [1957], p. 157: "It follows that if we have no information concerning the object of our observations, and the number of experiments n is not large, we cannot draw any conclusions; however, if n is sufficiently large, we shall obtain a good approximation with computations based on the assumption of a priori probabilities evenly distributed over all the possible values of the variable p."
[146] von Mises [1957], p. 116.
[147] von Mises [1957], p. 166. See also pp. vii, 159.
[148] See von Plato 1994, pp. 192–7.
[149] See Kendall 1939, pp. 87–9, who notes of von Mises's infinite series that "it is hardly surprising that this mystical concept has repelled a number of mathematicians and philosophers," before admitting "that von Mises' opponents have fared no better than he in providing a satisfactory basis for the theory of probability."

by mathematicians, who saw the definition as a concrete base on which a properly rigorous theory could be constructed. During the 1920s and 1930s, mathematicians such as Arthur Copeland and Abraham Wald developed various such approaches to probability. Some justified probability-as-frequency with the empirical observation of stable statistical ratios; others attempted to explain this stability from fundamental axioms. But all were based on the frequency interpretation.[150]

6.6.2. Andrei Kolmogorov's theory

The most successful of these mathematical theories was that of Andrei Kolmogorov, and was based on measure theory. Kolmogorov had started his studies of probability under Khintchine in the 1920s, and counted von Mises as his chief influence.[151] He defined probability as a measure property of a set within a field constructed according to a series of axioms. Though Kolmogorov published his first paper on the subject in Russian in 1929, his theory is usually dated to his great work *Grundbegriffe der Wahrscheinlichkeitsrechnung*, which appeared in 1933, the same year as the Jeffreys–Fisher dispute.[152]

Kolmogorov's theory was successful in the sense that current mathematical texts typically define a probability to be anything that satisfies Kolmogorov's axioms. Certainly his set theory approach was seen as the hallmark of mathematical rigor at the time; additionally, the concept of a random variable occurred naturally in his theory, rather than being artificially introduced as a primitive postulate as in von Mises's. Yet the retrospective claims of mathematicians such as J.L. Doob that the measure-theoretic approach came as a revelation, immediately displacing the statistical efforts of Fisher and Neyman, and establishing probability as a rigorous and thus respectable branch of pure mathematics, seem to be somewhat overstated.[153] Most texts on probability of the time are equivocal, none fully endorsing the measure-theoretic approach until after World War II. Von Plato points to affinities between Kolmogorov's work and studies of stochastic processes, and suggests that his theory was

---

[150] See von Plato 1994, pp. 192–7; also Fine 1973.

[151] During the 1960s, Kolmogorov attempted to revive von Mises's notion of randomness; see von Plato 1994, pp. 233–7.

[152] Kolmogorov 1933. Kolmogorov insisted his theory of probability should be viewed as an idealized system of mathematics, and not as a model of reality (a stance shared with von Mises). He thus always hedged his views on chance in nature. Von Plato 1994, pp. 204–6, 223, suggests that this was because the randomness Kolmogorov's theory enshrined contradicted the deterministic tenets of dialectical materialism.

[153] See, e.g., Doob 1989.

taken up most quickly by mathematicians who in the 1920s had been provoked by statistical physics to investigate randomness.[154] I will not pursue the matter, save to reiterate that alternatives to Kolmogorov's theory were also based on the frequency interpretation.

6.7. CONCLUSIONS

It would be wrong to pretend that the epistemic interpretation of probability was universally rejected during the 1930s. Philosophers, more concerned with logical structure than analysis of data, tended to persevere with the version made respectable by Keynes. W.E. Johnson, whose discussion of probabilistic hypotheses finally appeared in 1932, was influential in this regard.[155] His pupil Broad relied on the logical interpretation for his 1923 book *Scientific Thought*,[156] as did Jean Nicod for his 1930 *Foundations of Geometry and Induction*. Scientific philosophers of the 1930s such as Carnap and Popper also turned to Johnson.[157]

Nevertheless, even among philosophers the frequency model gained a purchase. The Cambridge-based Richard Braithwaite, interested in the scientific method, thought empirically observable frequencies the best approach to

[154] von Plato 1994, pp. 231–2.

[155] Johnson 1932. Johnson, who had died in 1931, wrote the three chapters published in *Mind* around 1925. They were intended to form the fourth part of his work *Logic*, the first three volumes of which had considered informal logic, syllogistic logic, and causality, respectively. Johnson's 'sufficientness postulate' eliminated the need to consider the probabilities of those outcomes that have not occurred, and his use of the concept of 'exchangeability' to generalize the Rule of Succession to the multinomial case obviated the balls-in-urn model. See Zabell 1982; Zabell 1989a, pp. 303–4, 305. De Finetti thought exchangeability more intuitive than statistical frequencies, and the concept is a key part of his later subjectivist development of probability.

[156] Referring again to the connection between epistemic probability and determinism, I note that most philosophers who wrote on the logic of probability were determinists. Broad 1920, p. 14, for example, was clear that the universe is rigidly causal and law-like: "if you find that some swans are white and that some are not, this is never the whole truth of the matter; all the white swans must have something common, and peculiar to them which 'accounts' for their whiteness." (Broad distanced himself slightly from this remark – it was merely a principle that "many people seem to believe" – but showed it to follow from his principle of uniformity.) It is our lack of knowledge of the correct causal categories – Mill's natural kinds – that results in uncertainty, and necessitates a probabilistic approach to induction.

[157] Carnap was one of the most prominent members of the Vienna School of logical positivism, and his later treatment of inductive logic as a branch of probability was a development of the positivists' attempt to reduce science to a series of statements that were 'meaningful' in the sense of being verifiably true or false. See Carnap 1950. Carnap termed the logical interpretation 'probability 1,' or the degree of 'confirmation,' and the frequency interpretation 'probability 2.'

probability in the exact sciences. Popper, who was skeptical of induction, rejected probability as a measure of the corroboration of a universal theory. But his focus on falsifiability as a hallmark of genuine scientific hypotheses, as laid out in his 1935 *Logic of Scientific Discovery*, is similar to Fisher's emphasis around the same time on the rejection of the null hypothesis.[158] (In *The Design of Experiments*, also of 1935, Fisher wrote: "Every experiment may be said to exist only in order to give the facts a chance of disproving the null hypothesis.[159])

As we have seen, probability assumed a central role in a number of academic disciplines during the 1930s. In each case it was the frequency interpretation that came to dominate, though for different reasons. Physicists took quantum mechanics to demonstrate that chance was inherent to the world, rather than an expression of uncertainty. Biologists also based their theories on the concept of randomness, though they were motivated by the lottery mechanism of population genetics rather than worries over causality. Social scientists, in contrast, were eager to win credibility through aping the old positivistic model of science, and shied from any hint of indeterminism. The model they embraced for experimentation and data analysis was, however, founded squarely on objective frequencies. Significantly, statistical primers written for social scientists are filled with examples from genetics. Statisticians too, in part responding to professional obligations, in part to the mathematization of the discipline, came to adopt the frequency interpretation.

---

[158] Popper's 'propensity' interpretation of probability was a half-baked attempt to apply the probabilities of von Mises's collectives to individual events.

[159] Fisher 1935a, p. 16.

7

Epilogue and Conclusions

7.1. EPILOGUE

At the start of the 1930s, the Bayesian position was regarded at best as a model of scientific inference, and as almost useless for analyzing data. Jeffreys was determined to remedy this situation, and from 1931 set about reproducing Fisher's frequentist methods of estimation and significance-testing from a Bayesian foundation. His *Theory of Probability* of 1939, the first book on applied Bayesian statistics, is a collection of such analytical tools. Many were derived specially for his seismological work. The method of 'bi-weighting,' for example, was developed to treat the systematic displacements and unusually large numbers of outliers that resulted when data of varying reliabilities from different seismic stations were combined. With Kenneth Bullen, one of his few graduate students, Jeffreys published the fruit of this work in 1940. The Jeffreys–Bullen tables, based on countless hours spent reducing and smoothing data, related average earthquake travel times to angular distances passed through the Earth.[1] Agreeing closely with Gutenberg's similar publication around the same time, the tables proved remarkably accurate in preserving genuine discontinuities in travel times, and are still used today to locate earthquake epicenters.

Geophysicists, however, remained more interested in Jeffreys's results than his methods; the few who thought they needed statistics at all believed least squares quite sufficient. Thus although Jeffreys maintained a steady stream of probabilistic publications after World War II, many on significance tests, his audience among physicists remained small.[2] Nor did statisticians pay much attention. The discipline had enjoyed a 'good war,' as engineers introduced

---

[1] Jeffreys and Bullen 1940.

[2] Personal contact with Jeffreys resulted in isolated pockets of influence. For instance, Bullen put the *Theory of Probability* on the reading list of the senior statistics course at the University of Sydney in the 1950s after arriving as Professor of Applied Mathematics. See Bolt 1991, p. 15.

statistical methods to streamline manufacturing processes, while the military promoted research toward enhancing artillery and bombing accuracy. The exigencies of wartime manufacture, together with the difficulties associated with items such as shells or bombs, which can only be tested destructively, also provoked a more sophisticated approach to sample surveys and reliability tests.[3] Military money supported research groups at several major American universities, including Neyman's at Berkeley, and these funds continued to flow after the end of hostilities.[4] The war thus served to enhance the status and professional autonomy of statistics, and to give statisticians new confidence in the relevance and importance of their discipline. Another consequence was a hardening of the opposition to inverse probability.[5] Much of the new research involved quality control, and grew from the decision theory developed by Neyman and Pearson during the 1930s. The aim of squeezing as much information as possible from a sample recalls the Bayesian approach, but the new methods were based firmly on a frequentist view of probability. The manufacturing process, for example, was modeled as repetitive and subject to random errors, and the issue was not about 'truth,' but whether, subject to the costs associated with Neyman and Pearson's two types of error, the process could be improved. Neyman's growing authority as a mathematician further strengthened the decision approach.[6]

During the 1940s, Bayesian approaches almost disappeared from mainstream mathematical statistics. Statisticians remained divided by doctrinal issues, but these stemmed from the ongoing battle between Neyman–Pearson and Fisherian schools. New graduate students shared these concerns. Dennis Lindley, a member of the burgeoning post-war statistics course at Cambridge, had little time for inverse probability. "We youngsters thought we knew a lot; we had been practicing statisticians for a few years and knew the proper way

---

[3] It was a military application that prompted Abraham Wald in 1943 to investigate whether more efficient tests could be found for experiments in which the final result became clear before the full sample was drawn. His 'sequential analysis' involves taking a decision whether to continue to sample after each element is tested; it can be considered a version of Neyman–Pearson with variable sample size, and results in a smaller average sampling size for a given risk of error. See Wald 1947. George Barnard was working on similar ideas in the U.K. around the same time.

[4] See Fienberg 1985; also Barnard and Plackett 1985.

[5] It is perhaps significant that unlike other statisticians and mathematicians working in Cambridge during the war, such as Wishart and G.H. Hardy, Jeffreys was apparently not informed about the cryptanalysis work at the Government Code and Cypher School at Bletchley Park.

[6] The wartime importance of statistics also introduced researchers in a number of academic disciplines to rule-driven numerical processes as models of rational decision-making. See Gigerenzer et al. 1989.

to proceed was by Fisher and Neyman–Pearson. (I had the curious idea that the two latter writers had done Fisher's mathematics correctly.) But Bayesian ideas, and Jeffreys: well, really, they were rather old hat and impractical. The need to take an examination, and perhaps the desire to sort out definitively what was wrong with Bayes, meant we went to the course."[7] Instead, it was the rigorous mathematical development of Bartlett's lectures that the young men wanted. Some had used statistics as part of war work, and now wanted to see it as "respectable a mathematical discipline as, say, analysis." Jeffreys's only research student in statistics, Vasant Huzurbazar, who arrived in Cambridge from India in 1946, recounts a similar experience. "The now respected phrase 'Bayesian inference' was not coined then, and instead the phrase 'Inverse Probability' was in use, but mostly with the intention of ultimate ridicule. Most of the books on statistics and the lengthy papers in journals in those days stressed repeatedly that they did not use inverse probability."[8]

Yet in the Introduction to this book I spoke not just of the fall but of the rise of inverse probability. From the mid-1950s Bayesianism, never completely extinct, started to re-emerge as a respectable alternative to the frequency view. The neo-Bayesian movement was heralded by the publication of I.J. Good's *Probability and the Weighing of Evidence* in 1950, and established with L.J. Savage's *Foundations of Statistics* four years later.[9] Immediately after the war, Good had often been the only member of the research section of the Royal Statistical Society to make Bayesian remarks; during the 1950s, however, an increasing number of statisticians began to propose Bayesian solutions to foundational problems. Dennis Lindley's conversion to the subjectivist position was influential, as was Howard Raffia and Robert Schlaifer's 1961 book *Applied Statistical Decision Theory*. Key debates of the period centered on the likelihood principle and the use of tail probabilities for statistical inference.

Bayesianism soon began to percolate into other disciplines. The first significant reappearance of the approach outside statistical circles seems to have been in psychology, in Ward Edwards's research group at the University of Michigan during the 1960s. This might appear strange, given my argument that Bayesianism's subjective nature made it an unattractive prospect for any

---

[7] Lindley 1980, pp. 35–6. Lindley, who converted to the Bayesian viewpoint during the 1950s, continues: "Not convinced, we had to admit that here was a cogent argument that even we rather arrogant young men could not demolish."

[8] Huzurbazar 1991, p. 19. See also Huzurbazar 1980. The American statistician Seymour Geisser recalls the Bayesian approach being regarded as "at worse, totally erroneous, at best, too restrictive and somewhere in between, outmoded." Geisser 1980, pp. 13, 16.

[9] Good 1950, 1965. Savage 1954. See also the collection of papers Good 1983.

but the most secure and self-confident of disciplines. But though Edwards – who had collaborated with Savage – tried to promote the approach for statistical analysis, the enduring element of his work was the use of Bayes's theorem as a testable model of human cognitive processes, in particular of intuition and learning in the light of new information. True, most psychologists who adopted this approach during the 1960s and 1970s tested it with conventional statistics. But the model of the mind as a Bayesian calculating machine initiated a lasting program of work in experimental psychology, and began to pervade other disciplines too.[10] This goes some way to explain why the Bayesian approach has to date made most headway in disciplines concerned with judgment and decision processes, such as economics and confirmation theory. Bayesianism has found less fertile ground in the physical sciences. Yet though the overwhelming majority of physicists continue to analyze their data using frequentist methods, a number of hard scientists such as E.T. Jaynes have proved persistent champions of Bayesian probability. In the last few years, mainstream textbooks have started to offer physicists Bayesian methods of analysis.[11]

With the revival in Bayesianism came Jeffreys's rehabilitation as a statistician. I.J. Good lauded the 1961 reprint of the *Theory of Probability* as "of greater importance for the philosophy of science, and obviously of greater immediate practical importance, than nearly all the books on probability written by professional philosophers *lumped together.*"[12] Two years later Jeffreys was awarded the Guy Medal of the Royal Statistical Society – despite having never attended a meeting – and the Cambridge Statistical Society hung his portrait. In 1980, the American economist Arnold Zellner organized a conference in Jeffreys's honor on Bayesian analysis in econometrics – a subject never of the remotest interest to the bemused honoree – and acclaimed his work as of vital importance for social scientists, and a source of inspiration too: "During the early years of this century he pressed on with his research on Bayesian philosophy and methods despite general dissent and indeed the force of his data and arguments has emerged triumphant."[13] The *Theory* was reprinted in paperback by Oxford University Press in 1983 following pressure from a wide group of probabilists and statisticians, and was reissued as part of the series "Oxford Classic Texts in the Physical Sciences" in 1998.

---

[10] See Gigerenzer et al. 1989, pp. 211–34.
[11] E.g., Sivia 1996. See also Jaynes 1990, Loredo 1990.
[12] Good 1962; see Good 1980, p. 32 for subsequent endorsement.
[13] Zellner 1980, p. 3. See also Jeffreys 1980.

My prime intention in writing this book has been to place the 1930s clash between the frequentist and epistemic interpretations within a more general overview of the history of probability, and to provide at the same time an account of the early work of R.A. Fisher and, especially, Harold Jeffreys. I have also tried to make the broader point that mathematical theories, like other products of science, can only be fully understood as products of their culture. The concept of probability is particularly pliable, and actors draw upon their experiences when seeking its interpretation and applications. For Fisher and Jeffreys, the meaning of probability and the development of a theory of inference evolved in each case with a particular conception of scientific practice, characterized by the differing aims and methods of genetics and geophysics respectively.

As indicated in the Introduction, many sorts of contextual resources can inform the understanding of chance and uncertainty. I have argued that though their nature and significance differed markedly between fields, it was disciplinary concerns that best account for the rise in frequentism and concomitant fall in Bayesianism during the 1930s. But one should not assume that the same will be true of all other periods. In the case of the post-war Bayesian revival, the precise nature of the factors at work remains an open question. Perhaps issues of race or gender will prove to have been significant, though I rather suspect not. But there is no reason to expect that a historical study of probability from 1945 would look quite as old-fashioned as my 'internal' account of the 1930s. For a start, it seems that the Bayesian revival owed much to the same wartime emphasis on action and behavior that served to strengthen the Neyman–Pearson position relative to Fisher's inferential approach. A number of neo-Bayesians came to the view through their search for a unified theory of acting, differing with Neyman–Pearson in relating the decision-making process directly to subjective preferences rather than via sampling frequencies. Others sought a coherent logic of induction, which the logical empiricists were failing to provide. It is also surely no coincidence that two of the more conspicuous proponents of the Bayesian view, Savage in the United States and Good in the U.K., worked on cryptanalysis and code-breaking technologies during the war, as assistants to von Neuman in Princeton and Alan Turing at Bletchley Park, respectively. (Savage's ideas on utility owe much to those of von Neumann and Oskar Morgenstern as developed in their work on game theory.) Perhaps the coherence and axiomatic foundations of the Bayesian approach – as opposed to the piecemeal nature of frequentist statistics – appealed to the prevailing academic enthusiasm for Grand Theory. It might even be not too far-fetched to link the rise of the subjective view during the 1950s and 1960s to the growing suspicion of received knowledge, especially from government sources, and the associated celebration of individual experience.

A full account of the Bayesian revival is still to be written. Yet one thing seems clear. The post-war context in which Bayesianism began to blossom differed significantly from the inimical environment of the 1930s. In consequence, the Bayesian position that emerged during the 1950s differed from what counted as a Bayesian view before the war. Neo-Bayesians such as Savage regarded as central the idea of 'personalistic' probability, and for the foundations of the subjective view turned not to Jeffreys but to the accounts of Bruno de Finetti or Frank Ramsey.[14] These writers insisted that a probability was based neither on frequencies nor on the logical relation between a proposition and a set of data, but instead represented nothing more than an individual's degree of confidence. As such, "two reasonable individuals faced with the same evidence may have different degrees of confidence in the truth of the same proposition."[15] In further contrast to Jeffreys, some personalists also held that subjective priors could only be partially ordered rather than sharply quantified.[16] The personalistic approach severed the uneasy bond between logical and subjective probabilities, and freed applications of the theory from the worries associated with prior probabilities. It could thus be framed in a more consistent and hence defensible way than the somewhat apologetic earlier versions of Keynes and Jeffreys. Those prepared to embrace the subjectivist view no longer had to claim that probability theory grounded a unique quantitative solution to the problem of induction.

I have argued that in the pre-war context at least, there was a natural affinity between frequency views of probability and indeterminism on the one hand and epistemic views and determinism on the other. De Finetti's work served

---

[14] De Finetti's 'representation theory' of 1928 allowed him to reduce objective frequencies – which he considered vague and metaphysical – to what he took to be their more fundamental subjective forms. See von Plato 1994, pp. 238–78. De Finetti published significant work on probability from the late 1920s, but it wasn't until he began to be championed by statisticians such as Savage in the mid 1950s that his work began to reach a wide audience, and was subsequently translated into English. (For de Finetti's contemporary gloss on Keynes and Jeffreys, see the reprinted review de Finetti 1985.) Ramsey developed his position in reaction to Keynes's claim that logical probabilities were accessible through intuition. Ramsey wrote that he could not perceive these probability relationships, and suspected that neither could anyone else. Like de Finetti, Ramsey became more influential in this post-war period than in the years immediately following the posthumous 1931 publication of his *The Foundations of Mathematics*. See Zabell 1991.
[15] Savage [1972], p. 3.
[16] See, e.g., Good 1980, pp. 24–5.

to break this simple association. Convinced that quantum mechanics had put paid to determinism, he had deliberately started work on his theory in order to sever the old ties between probability and determinism, believing that only a fully subjective view could be indifferent to causes of uncertainty, whether in the world or our heads. De Finetti's approach reconciled subjectivism with the indeterministic world-view of quantum mechanics. It also serves to support my argument that any connection drawn between such positions is not forced by logic, but is instead a contingent matter that seems in practice to depend on the intellectual environment of the person or group concerned. Though developing a view of probability similar to that of Jeffreys, de Finetti worked in a different intellectual context – in which the lessons of quantum physics bulked larger – and consequently interpreted his view of probability in a wholly different way.

These differing emphases explain why Fisher and Jeffreys, falling squarely into the opposed categories of 'frequentist' and 'Bayesian' at the time of their controversy, found themselves occupying a more ambiguous and fragmented space after the war. As mentioned, Jeffreys is now lionized as a heroic torch-carrier for Bayesianism during its wilderness years. Yet the reason his position was for many years "regarded as an eccentricity, with statisticians and prob-abilists alike ignoring it" was not solely because the *Theory of Probability* was too densely written, nor because – in I.J. Good's semi-serious sugges-tion – its lack of exercises for the student and overtly philosophical tone made it unsuitable as a course book, meaning no free copies for the instructor.[17] True, in the years immediately following the book's publication, the major-ity of statistical theorists were engaged in the dispute between confidence intervals and fiducial probability. But even the growing number sympathetic to the Bayesian cause during the 1950s found much to criticize in Jeffreys's approach. For example, stung in part by Fisher's objection that equivalent propositions would have different prior probabilities under transformation of their parameters, Jeffreys had devoted much effort after the war to the search

---

[17] Quote from Lindley 1991, p. 10; Good's suggestion from Good 1980, p. 22. (Admittedly, this is only one of several possible reasons advanced by Good.) A number of writers have commented on the style of the *Theory*. Seymour Geisser recalled "feeling that if one carefully scrutinized the work one could very likely find Bayesian solutions to most statistical problems – but exasperatingly, the search often seemed to require nearly as much effort as the research." Lindley amplified the point. "The *Theory* is a wonderfully rich book. Open it at almost any page, read carefully, and you will discover some pearl [but] Jeffreys's style does not give immediate comprehension. It is necessary to work at it. In my experience illumination usually appears and one wonders why it was so difficult to see at first. That is one reason why the book, although widely bought, has not been read or cited as much as it ought." See Geisser 1980, pp. 13–4; Lindley 1980, p. 37.

for an invariance theory of priors.[18] This attempt was not only unsuccessful, but stood diametrically opposed to the personalists' insistence that the assignment of prior probabilities was a matter for the individual, and not a necessary or indeed desirable part of a theory of probability. Jeffreys held that the probability of a proposition was fully specified given a set of data, and thus that while experience might be subjective, the use of the calculus of probability to generalize across experience was a perfectly objective process. This led to the irony that his work, dismissed by Fisher in the 1930s as "subjective and psychological," was accused of 'objectivism' by neo-Bayesians such as Savage during the 1950s and 1960s. Jeffreys's theory, they argued, missed completely the decision aspects that must be central to a theory specifically concerned with individual judgment. There is no mention of 'utility' in the *Theory*.[19] In a similar vein, philosophers such as Popper attacked Jeffreys's recommendation that prior probabilities should be assigned according to a universal standard of simplicity: the Simplicity Postulate was surely arbitrary, since 'simplicity' is a relative, not objective notion and depends on things like the coordinate system and the language of expression.[20] Some mathematical statisticians also sniped at his work as insufficiently rigorous.[21]

Fisher's position was uneasy too. Unique among his many innovations, fiducial probability won respect but few converts, and instead Neyman's confidence intervals came to dominate the statistical profession in the decade following the end of the war. Fisher continued to fight his corner with undiminished conviction and muscularity, but during the 1950s the substance of his argument began to change. He denied that all probabilities required direct

[18] See, for example, Jeffreys 1946. (His solution in this paper worked for a single unknown parameter, but broke down when extended to several considered jointly.) Jeffreys continued to search for a universal rule for prior probabilities for many years.

[19] See, e.g., Savage [1972], pp. 56–68. Jeffreys replied that Savage's objection – that people differ over their probability assessments – was precisely what had motivated the attempt to frame general rules for assigning prior probabilities. Such rules are not supposed to represent exact quantifications of ignorance; instead they are impartial working rules, necessary to ensure that the community of scientists can evaluate information in a clear and consistent way. (See, e.g., Jeffreys 1963.)

[20] See Howson 1988. Much of this criticism was unfair. Not only was Popper's belief that Jeffreys was purporting to justify induction mistaken, but Jeffreys had retracted some of his earlier claims for simplicity. See, e.g., Jeffreys 1936.

[21] See, for example, Good 1980, fn. 5, who recalled such a charge from H.A. Newman, then president of the London Mathematical Society, shortly before Jeffreys was knighted. In defense, Lindley 1980, p. 36, remarks that Jeffreys's work is "in the spirit of an applied mathematician of the Cambridge school which regards rigor as important in its proper place but never allows the study of it to prevent the basic principles being understood," and points out that Ramsey and de Finetti were often equally informal in their derivations.

interpretation as frequencies in a population. Some significance levels cannot be interpreted directly in terms of repeated random sampling, he argued, and instead the experimenter must use his judgment when fixing levels and treat each case on its merits. He also began to acknowledge that some condition of ignorance must be included in a problem's specification, and even conceded that his earlier claims that probabilities could not be attached to hypotheses were "hasty and erroneous."[22] The battle with Neyman forced Fisher much closer to Jeffreys's view of probability as a measure of subjective uncertainty. In concentrating on the scientific applications of statistical inference, rather than decisions or behavior based on that inference, Fisher and Jeffreys found themselves on the margins of both the non-inferential confidence interval and personalist approaches. One irascible, the other withdrawn, they became unlikely friends. Fisher credited Jeffreys's analysis of the Behrens–Fisher problem, and in their correspondence acknowledged the strengths of the Bayesian position. For his part, Jeffreys downplayed the earlier dispute, and took every opportunity to laud Fisher's intuitive brilliance, and to emphasize the similarity in their fundamental aims.

---

[22] See Zabell 1992, esp. pp. 379–80, for a discussion of Fisher and the fiducial argument. In *Statistical Methods and Scientific Inference* of 1956, p. 33, Fisher emphasized "the role of subjective ignorance, as well as that of objective knowledge in a typical probability statement," and, p. 59, went so far as to admit that perhaps fiducial and inverse probability statements were not so different after all: "It is essential to introduce the absence of knowledge *a priori* as a distinctive datum in order to demonstrate completely the applicability of the fiducial method of reasoning to the particular real and experimental cases for which it was developed. This point I failed to perceive when, in 1930, I first put forward the fiducial argument for calculating probabilities. For a time this led me to think that there was a difference in logical content between probability statements derived by different methods of reasoning. There are in reality no grounds for any such distinction."

Appendix 1

Sources for Chapter 2

My description of the development of probability theory from the mid-seventeenth to the end of the nineteenth centuries relies chiefly on secondary sources. Many of these were published in the last fifteen years. The literature of the history and philosophy of probability, though extremely rich, is still of manageable size.

Ian Hacking, in *The Emergence of Probability: a Philosophical Study of Early Ideas about Probability, Induction and Statistical Inference* (London: Cambridge University Press, 1975), describes how ideas of chance events and uncertain knowledge were blended during the late seventeenth and eighteenth centuries into a single theory of mathematical probability. In *Classical Probability in the Enlightenment* (Princeton, N.J.: Princeton University Press, 1988), Lorraine Daston examines the transformation of the classical calculus during the eighteenth century from a theory of games of chance to a model of rational thought. The new doctrine of chances was applied not only to the numerical uncertainty of insurance premiums or astronomical observations, but to matters of judgment in law, politics, and the moral sciences.

The nineteenth century is covered in Hacking's *The Taming of Chance* (Cambridge: Cambridge University Press, 1990), and Theodore Porter's *The Rise of Statistical Thinking, 1820–1900* (Princeton: Princeton University Press, 1986). Hacking argues that statistical and probabilistic methods were seen as granting a degree of predictability and control to the chaos of human behavior when applied in the social sphere during the early part of the century, but as undermining the doctrine of determinism when applied to the physical sciences towards the end. Porter describes the development of mathematical statistics in the social sciences, concentrating on the mutual influence between the frequency interpretation of probability and the perceived relationship between individual action and mass social phenomena.

Porter's *Trust in Numbers* (Princeton: Princeton University Press, 1995) can be regarded as a companion piece that examines, through the tensions between objective statistics and individual expertise, how this new form of numerical knowledge became authoritative. Donald MacKenzie's *Statistics in Britain, 1865–1930* (Edinburgh: Edinburgh University Press, 1981) is billed as an account of the social construction of scientific knowledge. He argues that the statistical techniques invented by the biometric school were bound up with their middle class values and advocacy of eugenic policy.

Many of the more mathematical texts are also sensitive to the dependence of statistics and probability on the cultural and philosophical context. Stephen Stigler's *The History of Statistics: The Measurement of Uncertainty before 1900* (Cambridge: Belknap Press of Harvard University Press, 1986) is the definitive study of statistical measurement during the nineteenth century. Anders Hald covers an earlier period in his *A History of Probability and Statistics and their Applications before 1750* (New York: Wiley, 1990), and focuses on the application of probabilistic inference to a number of problems in gambling, life insurance, and the analysis of astronomical errors. F.N. David's delightful *Games, Gods and Gambling: A History of Probabilistic and Statistical Ideas* (London: Charles Griffin, 1962) also covers this early history of the subject. *Creating Modern Probability: Its Mathematics, Physics, and Philosophy in Historical Perspective* (Cambridge: Cambridge University Press, 1994) by Jan von Plato shows how the various interpretations of probability were transformed in the first half of the twentieth century into branches of pure mathematics. Andrew Dale concentrates on the equations in his indispensable *A History of Inverse Probability: From Thomas Bayes to Karl Pearson* (New York: Springer-Verlag, 1991). Both Isaac Todhunter's classic *History of the Mathematical Theory of Probability* (London: Macmillan, 1865) and John Maynard Keynes's stylish *Treatise on Probability* (London: Macmillan, 1921), though primary sources in their own right, remain essential surveys of the field. The same can be said of Karl Pearson's endlessly rewarding *The History of Statistics in the 17th and 18th Centuries*, ed. E.S. Pearson (London: Charles Griffin, 1978).

There are also a number of useful shorter studies of the subject. Daston, Hacking, Porter, von Plato, and Stigler were among those who attended a year-long research project in the history and philosophy of probabilistic thought at the Zentrum für interdisziplinäre Forschung at the University of Bielefeld in West Germany from 1982 to 1983. A number of essays arising from this project are collected into the two-volume *The Probabilistic Revolution* (Cambridge: MIT Press, 1987). The first volume, edited by Lorenz Krüger, Daston, and Michael Heidelberger, is subtitled "Ideas in History"; the second, edited by Krüger, Gerd Gigerenzer, and Mary Morgan, "Ideas in the Sciences." Some of the same group have produced a more coherent account of the dominant themes. *The Empire of Chance* (Cambridge: C.U.P., 1989) is co-authored by Gigerenzer, Zeno Swijtink, Porter, Daston, John Beatty, and Krüger. ("Dutiful subjects of the empire of chance, we used a lottery to order our names on the title page.") Egon Pearson and Maurice Kendall are the editors of a collection of essays on historical themes by mathematicians and statisticians, *Studies in the History of Statistics and Probability* (London: Charles Griffin, 1970). A second volume appeared in 1977, edited by Kendall with Robin Plackett: *Studies in the History of Statistics and Probability: Vol. 2* (London: Charles Griffin, 1977). The brief *Historical Aspects of the Bayesian Controversy* by J. Weber (Tucson, Arizona: Division of Economic and Business Research, 1973) is also worth a look. Finally, a number of articles in the two-volume *Companion Encyclopedia of the History and Philosophy of the Mathematical Sciences* (London: Routledge, 1994), edited by Ivor Grattan-Guinness, deal with probabilistic and statistical methods and their applications in the physical and social sciences.

Appendix 2

Bayesian Conditioning as a Model of Scientific Inference

Consider a hypothesis, h, and an experimental result, e. Bayes's theorem states that

$$P(h|e) = P(e|h)P(h)/P(e)$$

The conditional probability of h given e can be interpreted as the degree of rational belief in the hypothesis given the evidence. Bayes's theorem therefore governs the evaluation of hypotheses in the light of new data, and can be regarded as a model of the scientific method.

The model seems to fit with several features of observed practice. The experimental result will support the hypothesis if $P(h|e) > P(h)$, undermine it if $P(h|e) < P(h)$, and be irrelevant if $P(h|e) = P(h)$. From Bayes's theorem, the result will increase the probability of a given hypothesis in proportion to its likelihood, $P(e|h)$. So if the evidence is impossible on the hypothesis, $P(e|h) = 0$, then $P(h|e) = 0$ too, and the hypothesis is refuted, as required. No amount of evidence can ever wholly confirm a hypothesis, but if the evidence is entailed by the hypothesis, $P(e|h) = 1$, and its observation provides maximum support.

The posterior probability of the hypothesis is proportional to its prior probability. Small quantities of consistent evidence will therefore not make plausible an initially improbable hypothesis. (Unless this hypothesis is the only way to account for the evidence; like Sherlock Holmes, Bayes's theorem indicates that after eliminating the impossible, whatever remains, however improbable, must be the truth.) Conversely, a hypothesis initially regarded with near certainty would not be shaken by the occasional observation of an improbable result.

The change in posterior probability of a hypothesis following a new result also depends inversely on the intrinsic probability of that result. Hence the intuition that surprising or unusual evidence has a greater effect in raising the probability of a hypothesis than a commonplace.

Finally, Bayes's theorem can account for scientists' reluctance to discard wholesale a complex but successful theory in the face of a contradictory result. Dorling has illustrated this with a "quasi-historical numerical example" – Adams's nineteenth-century explanation of the Moon's secular acceleration.*

* See Dorling 1979, Redhead 1980; also Howson and Urbach 1993, pp. 117–164. For more on Bayesianism see Press 1989, Howson and Urbach 1993, Earman 1992.

Appendix 3

Abbreviations Used in the Footnotes

GAB   Private correspondence: DH and Professor George Barnard.

MSB   Private correspondence: DH and Professor M.S. Bartlett.

IJG   Interview: DH and I.J. Good, audio-tape (11 December 1996; corrected transcript, February 27 1998).

BJ   Interview: DH and Bertha Swirles, Lady Jeffreys, audio-tape; A.I.P. Collection (March 1997).

HJ   Archive: Jeffreys Papers.

HJ-GAB   Interview: Harold Jeffreys and G.A. Barnard, audio-tape (13 May 1982).

HJ-DVL   Interview: Harold Jeffreys and D.V. Lindley, video-tape (25 August 1983; corrected transcript, 3 November 1986).

HJ-MM   Interview: Harold Jeffreys and Michael McIntyre (16 September 1984).

Bibliography

P.G. Abir-Am, "Synergy or clash: disciplinary and marital strategies in the career of mathematical biologist Dorothy Wrinch," in P.G. Abir-Am and D. Outram, eds., *Uneasy Careers and Intimate Lives: Women in Science, 1789-1979* (New Brunswick, N.J., Rutgers Univ. Press, 1987).

P.G. Abir-Am, "Dorothy Maud Wrinch (1894-1976)," in L.S. Grinstein et al., eds., *Women in Chemistry and Physics* (New York: Greenwood, 1993).

K. Alder, "A revolution to measure: the political economy of the metric system in France," in Wise, 1995.

J. Aldrich, *R.A. Fisher and the Making of Maximum Likelihood, 1912-22* (Department of Economics, University of Southampton, 1995).

G. Allen, *Life Science in the Twentieth Century* (New York: Wiley, 1975).

A.J. Ayer, "Still more of my life," in *The Philosophy of A.J. Ayer*, ed. L.E. Hahn (La Salle, IL: Open Court, 1992).

G.A. Barnard, "Fisher's contributions to mathematical statistics," J. Roy. Stat. Soc. A **126**, 162-166 (1963).

G.A. Barnard, "R.A. Fisher – a true Bayesian?" Internat. Stat. Rev. **55**, 183-189 (1987).

G.A. Barnard, Reply to S.L. Zabell, Stat. Sci. **4**, 258-260 (1989).

G.A. Barnard, "Fisher: a retrospective," Chance **3**, 22-28 (1990).

G.A. Barnard, "Neyman and the Behrens-Fisher problem – an anecdote," Prob. and Math. Stat. **15**, 67-71 (1995).

G.A. Barnard and R.L. Plackett, "Statistics in the United Kingdom, 1939-45," in A.C. Atkinson and S.E. Fienberg, eds., *A Celebration of Statistics": the ISI Centenary Volume* (New York: Springer, 1985).

B. Barnes, D. Bloor, and J. Henry, *Scientific Knowledge: a Sociological Analysis* (Chicago: U.C.P., 1996).

M.S. Bartlett, "Probability and chance in the theory of statistics," Proc. Roy. Soc. A **141**, 518-534 (1933).

M.S. Bartlett, "The present position of mathematical statistics," Jou. Roy. Stat. Soc. **103**, 1-29 (1940).

M.S. Bartlett, "R.A. Fisher and the first fifty years of statistical methodology," Jou. Am. Stat. Ass. **60**, 395-409 (1965).

M.S. Bartlett, "Chance and change," in *The Making of Statisticians*, ed., J. Gani (New York: Springer-Verlag, 1982); reprinted in Bartlett 1989.

M.S. Bartlett, *Selected Papers of M.S. Bartlett*, 3 vols. (Winnipeg, Canada: Charles Babbage Research Centre, 1989).

T. Bayes, "An essay towards solving a problem in the doctrine of chances," Phil. Trans. Roy. Soc. Lon. **53**, 370-418 (1764); reprinted in Pearson and Kendall, 1970.

C. Bazerman, *Shaping Written Knowledge: The Genre and Activity of the Experimental Article in Science* (Madison: Univ. Wisc. Press, 1988).

J.H. Bennett, ed., *Natural Selection, Heredity, and Eugenics: including Selected Correspondence of R.A. Fisher with Leonard Darwin and Others* (Oxford: O.U.P., 1983).

J.H. Bennett, ed., *Statistical Inference and Analysis: Selected Correspondence of R.A. Fisher* (Oxford: O.U.P., 1990).

P.C. Bhat, H.B. Prosper, and S.S. Snyder, "Bayesian analysis of multi-source data," Phys. Lett. B **407**, 73-8 (1997).

J. Bibby, *Notes Towards a History of Teaching Statistics* (Edinburgh: John Bibby, 1986).

D. Bloor, *Science and Social Imagery* (Chicago: University of Chicago Press, 1976; second edition, 1991).

B.A. Bolt, "Sir Harold Jeffreys and geophysical inverse problems," Chance **4**, 15-17 (1991).

L.C.W. Bonacina, "Relativity, space, and ultimate reality"; " 'Space' or 'aether' "; "The physical continuity of 'space,' " Nature **107**, 171-2; 234; 300 (1921).

P.J. Bowler, *Evolution: the history of an idea* (Berkeley: University of California Press, 1984).

P.J. Bowler, *The Eclipse of Darwinism: Anti-Darwinism Evolution Theories in the Decades Around 1900* (Baltimore: Johns Hopkins Univ. Press, 1992).

J.F. Box, *R.A. Fisher: the Life of a Scientist* (New York: Wiley, 1978).

J. van Brakel, "The possible influence of the discovery of radio-active decay on the concept of physical probability," Arch. Hist. Exact Sci. **31**, 369-85 (1985).

C.D. Broad, "On the relation between induction and probability (part 1)," Mind **27**, 389-404 (1918).

C.D. Broad, "On the relation between induction and probability (part 2)," Mind **29**, 11-45 (1920).

C.D. Broad, Review of Keynes's *A Treatise on Probability*, Mind **31**, 72-85 (1922).

J. Bromberg, *The Laser in America* (Cambridge, Mass.: M.I.T. Press, 1991).

S. Brush, "Prediction and theory evaluation: the case of light bending," Science **246**, 1124-29 (1989).

S. Brush, *Nebulous Earth: the Origin of the Solar System and the Core of the Earth from Laplace to Jeffreys* (Cambridge: C.U.P., 1996).

J. Buchwald, *The Creation of Scientific Effects: Heinrich Hertz and electric waves* (Chicago: University of Chicago Press, 1994).

W.H.L.E. Bulwer, *France: Social, Literary, Political* (London, 1834).

L. Cameron and J. Forrester, "Tansley's psychoanalytic network: An episode out of the early history of psychoanalysis in England," Psychoanalysis and History **2**, 189-256 (2000).

N.R. Campbell, *Physics: the Elements* (Cambridge: C.U.P., 1919).

A. Carabelli, *On Keynes's Method* (London: Macmillan, 1988).

R. Carnap, *Logical Foundations of Probability* (Chicago: University of Chicago Press, 1950).

N. Cartwright, "Response of American physicists," in Krüger, Gigerenzer, and Morgan 1987.

A. Church, "On the concept of a random sequence," Bull. Am. Math. Soc. **46**, 130-135 (1940).

W.G. Cochran, Review of Van Deuren's *La Théorie de Probabilités*, Jou. Roy. Stat. Soc. **99**, 169-70 (1936).

W. Coleman, *Biology in the Nineteenth Century* (New York: Wiley, 1971)

H. Collins, *Changing Order: Replication and Induction in Scientific Practice* (London: Sage, 1985).

D. Conniffe, "Keynes on probability and statistical inference and the links to Fisher," Cam. Jou. Econ. **16**, 475-89 (1992).

L.R. Connor, Review of J.B. Scarborough's *Numerical Mathematical Analysis*, Jou. Roy. Stat. Soc. **94**, 442-3 (1931).

J.M. Converse, *Survey Research in the United States, 1890-1960* (Berkeley, CA: University of California Press, 1987).

A. Cook, "Sir Harold Jeffreys," Biog. Mem. Fell. Roy. Soc. **36**, 301-333 (1990).

J.L. Coolidge, *An Introduction to Mathematical Probability* (Oxford: O.U.P., 1925).

A. Cottrell, "Keynes's theory of probability and its relevance to his economics," Econ. Phil. **9**, 25-51 (1993).

A.I. Dale, *A History of Inverse Probability: From Thomas Bayes to Karl Pearson* (New York: Springer-Verlag, 1991).

K. Danziger, "Research practice in American psychology," in Krüger, Gigerenzer, and Morgan, 1987.

L. Daston, "D'Alembert's critique of probability theory," Historia Mathematica **6**, 259-79 (1979).

L. Daston, "The domestication of risk: mathematical probability and insurance, 1650-1830," in L. Krüger, L. Daston, and M. Heidelberger, 1987.

L. Daston, *Classical Probability in the Enlightenment* (Princeton: Princeton University Press, 1988).

L. Daston and P. Galison, "The image of objectivity," Representations **40**, 81-128 (1992).

F.N. David, *Games, Gods and Gambling: A History of Probabilistic and Statistical Ideas* (London: Charles Griffin, 1962).

A. Desmond and J. Moore, *Darwin* (New York: Warner Books, 1991).

H. Dingle, *Science and Human Experience* (London: Williams and Norgate, 1931).

H. Dingle, *Through Science to Philosophy* (Oxford: Clarendon Press, 1937).

P.A.M. Dirac, *Principles of Quantum Mechanics* (Oxford: Clarendon Press, 1930).

J.L. Doob, "Kolmogorov's early work on convergence theory and foundations," Annals of Probability **17**, 815-21 (1989).

J. Dorling, "Bayesian personalism, the methodology of scientific research programmes, and Duhem's problem," Stud. Hist. Phil. Sci. **10**, 177-187 (1979).

J. Earman, *Bayes or Bust?* (Cambridge, Mass.: M.I.T. Press, 1992).

A. Eddington, "Notes on the method of least squares," Proc. Phys. Soc. **45**, 271-282 (1933).

A. Eddington, *New Pathways in Science* (Cambridge: C.U.P., 1935).

A. Eddington, *The Philosophy of Physical Science* (Cambridge: C.U.P., 1939).

F.Y. Edgeworth, "Applications of probabilities to economics," Econ. Jou. **20**, 284-301; 441-465 (1910).

A.W.F. Edwards, *Likelihood* (Cambridge: C.U.P., 1972; revised ed. Johns Hopkins, 1992).

A.W.F. Edwards, "The history of likelihood," Int. Stat. Rev. **42**, 9-15 (1974).

A.W.F. Edwards, "Fiducial probability," The Statistician, **25**, 15-35 (1976).

W. M. Elsasser, *Memoirs of a Physicist in the Atomic Age* (New York: Science History Publications, 1978.)

R.A. Epstein, *The Theory of Gambling and Statistical Logic* (San Diego, CA: Academic Press, 1967 [1977]).

S.E. Fienberg, "Statistical developments in World War II: an international perspective," in A.C. Atkinson and S.E. Fienberg, eds., *A Celebration of Statistics: the ISI Centenary Volume* (New York: Springer, 1985).

T.L. Fine, *Theories of Probability* (New York: Academic Press, 1973).

B. de Finetti, "Cambridge probability theorists," Manchester School **53**, 347-363 (1985).

R.A. Fisher, "On an absolute criterion for fitting frequency curves," Messeng. Math. **41**, 155-60 (1912).

R.A. Fisher, "Some hopes of a eugenist," Eugenics Rev. **5**, 309-15 (1914).

R.A. Fisher, "Frequency distribution of the values of the correlation coefficient in samples from an indefinitely large population," Biometrika **10**, 507-21 (1915).

R.A. Fisher, "The correlation between relative on the supposition of Mendelian inheritance," Trans. Roy. Soc. Edin. **52**, 399-433 (1918).

R.A. Fisher, "On the probable error of a coefficient of correlation deduced from a small sample," Metron **1**, 1-32 (1921).

R.A. Fisher, "On the mathematical foundations of theoretical statistics," Phil. Trans. Roy. Soc. Lon. A **222**, 309-68 (1922a).

R.A. Fisher, Review of Keynes's *A Treatise on Probability*, Eugenics Rev. **14**, 46-50 (1922b).

R.A. Fisher, "Theory of statistical estimation," Proc. Cam. Phil. Soc. **22**, 700-25 (1925a).

R.A. Fisher, *Statistical Methods for Research Workers* (Edinburgh: Oliver and Boyd, 1925b).

R.A. Fisher, "The arrangement of field experiments," Jou. Ministry Agriculture Great Britain **33**, 503-513 (1926).

R.A. Fisher, "Inverse probability," Proc. Cam. Phil. Soc. **26**, 528-535 (1930a).

R.A. Fisher, *The Genetic Theory of Natural Selection* (Oxford: Clarendon, 1930b).

R.A. Fisher, "Inverse probability and the use of likelihood," Proc. Cam. Phil. Soc. **28**, 257-261 (1932).

R.A. Fisher, "The concepts of inverse probability and fiducial probability referring to unknown parameters," Proc. Roy. Soc. A **139**, 343-348 (1933).

R.A. Fisher, "Probability likelihood and quantity of information in the logical of uncertain inference," Proc. Roy. Soc. A **146**, 1-8 (1934).

R.A. Fisher, *The Design of Experiments* (Edinburgh: Oliver and Boyd, 1935a).

R.A. Fisher, "The Logic of Inductive Inference," Jou. Roy. Stat. Soc. A **98**, 39-82 (1935b).

R.A. Fisher, "The fiducial argument in statistical inference," Annals Eugenics **6**, 391-398 (1935c)

R.A. Fisher, "Uncertain inference," Proc. Amer. Acad. Arts Sci. **71**, 245-258 (1936).

R.A. Fisher, "Student," Annals Eugenics **9**, 1-9 (1939).

R.A. Fisher, *Statistical Methods and Scientific Inference* (Edinburgh: Oliver and Boyd, 1956).

R.A. Fisher, "The nature of probability," Centennial Review **2**, 261-274 (1958).

P. Forman, "Weimar culture, causality, and quantum theory, 1918-1927: Adaptation by German physicists and mathematicians to a hostile intellectual milieu," Hist. Stud. Phys. Sci. **3**, 1-115 (1971).

P. Forman, "The reception of an acausal quantum mechanics in Germany and Britain," in *The Reception of Unconventional Science*, ed. S.H. Mauskopf (Washington, D.C.: American Association for the Advancement of Science, 1979).

P. Forman, "Inventing the maser in postwar America," Osiris **7**, 105-134 (1992).

D.W. Forrest, *Francis Galton: the Life and Works of a Victorian Genius* (New York: Taplinger, 1974).

P. Galison, *How Experiments End* (Chicago: 1987).

F. Galton, *Natural Inheritance* (London: Macmillan, 1889).

S. Geisser, "The contributions of Sir Harold Jeffreys to Bayesian inference," in A. Zellner, ed., *Bayesian Analysis in Econometrics and Statistics* (Amsterdam: North Holland, 1980).

G. Gigerenzer, "Probabilistic thinking and the flight against subjectivity," in Krüger, Gigerenzer, and Morgan, 1987.

G. Gigerenzer, Z. Swijtink, T. Porter, L. Daston, J. Beatty, and L. Krüger, *The Empire of Chance* (Cambridge: C.U.P., 1989).

R. Glenday, "The use and measure of economic statistics," Jou. Roy. Stat. Soc. **98**, 497-505 (1935).

I.J. Good, *Probability and the Weighing of Evidence* (London: Charles Griffin, 1950).

I.J. Good, Review of Jeffreys's Theory of Probability, Jou. Roy. Stat. Soc. A **125**, 487-489 (1962).

I.J. Good, *The Estimation of Probabilities* (Cambridge, Mass.: M.I.T. Press, 1965).

I.J. Good, "The contributions of Jeffreys to Bayesian statistics," in A. Zellner, ed., *Bayesian Analysis in Econometrics and Statistics* (Amsterdam: North Holland, 1980).

I.J. Good, *Good Thinking* (Minneapolis: University of Minnesota Press, 1983).

G. Gooday, "The morals of energy metering: constructing and deconstructing the precision of the Victorian electrical engineer's ammeter and voltmeter," in M. Wise 1995.

A.P.L. Gordon, "The statistics of totalisator betting," Jou. Roy. Stat. Soc. **94**, 31-76 (1931).

S.J. Gould, *The Panda's Thumb* (New York: Norton, 1980).

I. Grattan-Guinness, ed., *Companion Encyclopedia of the History and Philosophy of the Mathematical Sciences* (London : Routledge, 1994).

M. Greenwood, Review of H. Westergaard's *Contributions to the History of Statistics*, Jou. Roy. Stat. Soc. **95**, 720-2 (1932).

M. Greenwood, "On the value of Royal Commissions in sociological research, with special reference to the birth-rate," Jou. Roy. Stat. Soc. **100**, 396-401 (1937).

G.R. Grimmett and D.R. Stirzaker, *Probability and Random Processes* (Oxford: O.U.P., 1982 [1992]).

I. Hacking, *The Emergence of Probability : a Philosophical Study of Early Ideas about Probability, Induction and Statistical Inference* (London: C.U.P., 1975)

I. Hacking, *The Taming of Chance* (Cambridge: C.U.P., 1990).

I. Hacking, "The self-vindication of the laboratory sciences," in A. Pickering, ed., *Science as Practice and Culture* (Chicago: U.C.P., 1992).

A. Hald, *A History of Probability and Statistics and their Applications before 1750* (New York: Wiley, 1990).

J.B.S. Haldane, "A note on inverse probability," Proc. Cam. Phil. Soc. **28**, 55-61 (1932a).

J.B.S. Haldane, "Determinism," Nature **129**, 315-6 (1932b).

J.B.S. Haldane, "Karl Pearson, 1857-1957," Biometrika **44**, 303-313 (1957); reprinted in Pearson and Kendall, 1970.

M.H. Hansen and W.G. Madow, "Some important events in the historical development of sample surveys," in Owen, 1976.

B. Harshbarger, "History of the early developments of modern statistics in America (1920-44)," in Owen, 1976.

J. Heilbron, "The earliest missionaries of the Copenhagen spirit," in E. Ullmann-Margalit, ed., *Science in Reflection*, BSPS **110** (Dordrecht: Kluwer, 1988).

J. Heilbron and R. Seidel, *Lawrence and His Laboratory* (Berkeley: University of California Press, 1989).

J. Hendry, "Weimar culture and quantum causality," Hist. Sci. **18**, 155-180 (1980).

A.B. Hill, "The enumeration of population of the meetings of the society," Jou. Roy. Stat. Soc. **99**, 162-164B (1936).

I.D. Hill, "Statistical Society of London – Royal Statistical Society, the first 100 years: 1834-1934," Jou. Roy. Stat. Soc. A **147**, 130-9 (1984).

B.K. Holland, "The Chevalier de Méré and the power of excessive gambling," Chance **4**, 5 (1991).

A. Holmes, *The Age of the Earth* (London: Harper, 1913).

R.A. Horváth, "The rise of macroeconomic calculations in economic statistics," in Krüger, Gigerenzer, and Morgan, 1987.

H. Hotelling, Review of Fisher's *Statistical Methods for Research Workers*, Jou. Am. Stat. Soc. **22**, 411-2 (1927).

B.F. Howell, *An Introduction to Seismological Research: History and Development* (Cambridge: C.U.P., 1990).

C. Howson, "On the consistency of Jeffreys's Simplicity Postulate, and its role in Bayesian inference," Phil. Quart. **38**, 68-83 (1988).

C. Howson and P. Urbach, *Scientific Reasoning: The Bayesian Approach* (Chicago: Open Court, 1993).

B. Hunt, "Rigorous discipline: Oliver Heaviside versus the mathematicians," in P. Dear, ed., *The Literary Structure of Scientific Argument: Historical studies* (Philadelphia: University of Pennsylvania Press, 1991a).

B. Hunt, *The Maxwellians* (Ithaca: Cornell University Press, 1991b).

V.S. Huzurbazar, "Bayesian inference and invariant prior probabilities," in A. Zellner, ed., *Bayesian Analysis in Econometrics and Statistics* (Amsterdam: North Holland, 1980).

V.S. Huzurbazar, "Sir Harold Jeffreys: recollections of a student," Chance **4**, 18-21 (1991).

J.O. Irwin, "Recent advances in mathematical statistics," Jou. Roy. Stat. Soc. **97**, 114-54 (1934).

J.O. Irwin (with W.G. Cochran, E.C. Fieller, and W.L. Stevens), "Recent advances in mathematical statistics," Jou. Roy. Stat. Soc. **99**, 714-69 (1936).

J.O. Irwin, Review of Jeffreys's *Theory of Probability*, Jou. Roy. Stat. Soc. **104**, 59-64 (1941).

L. Isserlis, "Inverse probability," Jou. Roy. Stat. Soc. **99**, 130-7 (1936).

E. Jaynes, "Probability theory as logic," in P. Fougère, ed., *Maximum Entropy and Bayesian Methods* (Dordrecht: Kluwer, 1990).

B. Jeffreys, "Harold Jeffreys: some reminiscences," Chance **4**, 22-23, 26 (1991).

B. Jeffreys, "Reminiscences and discoveries: Harold Jeffreys from 1891 to 1940," Notes. Rec. R. Soc. Lon. **46**, 301-8 (1992).

H. Jeffreys, "On the crucial tests of Einstein's theory of gravitation," Mon. Not. Roy. Astron. Soc. **80**, 138-154 (1919).

H. Jeffreys, "Tidal friction in shallow seas," Phil. Trans. Roy. Soc. Lon. A **221**, 239-64 (1920).

H. Jeffreys, "The concept of space in physics," Nature **107**, 267-8; "The physical status of space," Nature **107**, 394 (1921a).

H. Jeffreys, "Uniform motion in the aether," Nature **107**, 7467; **108**, 80 (1921b).

H. Jeffreys, "The age of the Earth," Nature **108**, 284, 370 (1921c).

H. Jeffreys, "Relativity and materialism," Nature **108**, 568-9 (1921d).

H. Jeffreys, Review of Keynes's *A Treatise on Probability*, Nature **109**, 132-3 (1922).

H. Jeffreys, "The Pamir earthquake of 1911 Feb. 18 in relation to the depths of earthquake foci," Mon. Not. Roy. Astron. Soc. Geophys. Suppl. **1**, 22-31 (1923a).

H. Jeffreys, Review of Campbell's *Physics, the Elements*, Phil. Mag. **46**, 1021-5 (1923b).

H. Jeffreys, *The Earth: Its Origin, History, and Physical Constitution* (Cambridge: C.U.P., 1924).

H. Jeffreys, "The evolution of the Earth as a planet," in *Evolution in the Light of Modern Knowledge* (London: Blackie and Son, 1925).

H. Jeffreys, "The rigidity of the Earth's central core," Mon. Not. Roy. Astron. Soc. Geophys. Suppl. **1**, 371-83 (1926).

H. Jeffreys, *Operational Methods in Mathematical Physics* (Cambridge: C.U.P., 1927).

H. Jeffreys, Review of Bridgman's *The Logic of Modern Physics*, Nature **121**, 86-7 (1928).

H. Jeffreys, "The planetismal hypothesis," Observatory **52**, 173-7 (1929).

H. Jeffreys, *Scientific Inference* (Cambridge: C.U.P., 1931; reissued with addenda, 1937).

H. Jeffreys, "An Alternative to the Rejection of Observations," Proc. Roy. Soc. A **137**, 78-87 (1932a).

H. Jeffreys, "On the theory of errors and least squares," Proc. Roy. Soc. A **138**, 48-54 (1932b).

H. Jeffreys, "On the prior probability in the theory of sampling," Proc. Cam. Phil. Soc. **29**, 83-7 (1932c).

H. Jeffreys, "Probability, statistics, and the theory of errors," Proc. Roy. Soc. A **140**, 523-535 (1933a).

H. Jeffreys, "On Gauss's proof of the normal law of errors," Proc. Cam. Phil. Soc. **29**, 231-4 (1933b).

H. Jeffreys, "Probability and scientific method," Proc. Roy. Soc. A **146**, 9-16 (1934).

H. Jeffreys, "On some criticisms of the theory of probability," Phil. Mag. **22**, 337-359 (1936).

H. Jeffreys, "Modern Aristotelianism: contribution to discussion," Nature **139**, 1004-5 (1937).

H. Jeffreys and K.E. Bullen, *Seismological Tables* (British Association: the Gray-Milne Trust, 1940).

H. Jeffreys, "Probability and quantum theory," Phil. Mag. **33**, 815-31 (1942).

H. Jeffreys, "An invariant form of the prior probability in estimation problems," Proc. Roy. Soc. A **186**, 453-61 (1946).

H. Jeffreys, *Theory of Probability* (Oxford: Clarendon Press, 1939; second edition 1948; third edition 1961).

H. Jeffreys, Review of L.J. Savage, ed., *The Foundations of Statistical Inference*, Technometrics **5**, 407-10 (1963).

H. Jeffreys, "Fisher and inverse probability," Int. Stat. Rev. **42**, 1-3 (1974).

H. Jeffreys, "Some general points in probability theory," in A. Zellner, ed., *Bayesian Analysis in Econometrics and Statistics* (Amsterdam: North Holland, 1980).

W.E. Johnson, "Probability," Mind **41**, 1-16; 281-96; 409-23 (1932).

D.C. Jones and C.G. Clark, "Housing in Liverpool: a survey by sample of present conditions," Jou. Roy. Stat. Soc. **93**, 489-537 (1930).

G. Jorland, "The Saint Petersburg paradox," in L. Krüger, L. Daston, and M. Heidelberger, 1987.

M.G. Kendall and B.B. Smith, "Randomness and random sampling numbers," Jou. Roy. Stat. Soc. **101**, 147-66 (1938).

M.G. Kendall, Review of von Mises's *Wahrscheinlichkeit, Statistik, und Wahrheit*, Jou. Roy. Stat. Soc. **102**, 87-9 (1939).

M.G. Kendall, "George Udny Yule, 1871-1951," Jou. Roy. Stat. Soc. **115A**, 156-61 (1952); reprinted in Pearson and Kendall, 1970.

M.G. Kendall and R.L. Plackett, eds., *Studies in the History of Statistics and Probability: Vol. 2* (London: Griffin, 1977).

Kennet of the Dere, Lord [i.e., E.H. Young], "The consumption of statistics," Jou. Roy. Stat. Soc. **100**, 1-17 (1937).

D. Kevles, *In the Name of Eugenics* (Berkeley: University of California Press, 1985).

J.M. Keynes, *Treatise on Probability* (London: Macmillan, 1921; reprinted 1973).

J.M. Keynes, *Essays in Biography* (London: Macmillan, 1933; revised edition, ed. G. Keynes, New York: Norton, 1963).

J.M. Keynes, *Two Memoirs: Dr Melchior, a Defeated Enemy, and My Early Beliefs* (London: Rupert Hart-Davis, 1949).

L. Knopoff, "Sir Harold Jeffreys: *The Earth: Its Origin, History, and Physical Constitution*," Chance **4**, 24-26 (1991).

A. Koestler, *The Case of the Mid-Wife Toad* (London: Hutchinson, 1971).

R.E. Kohler, *Lords of the Fly* (Chicago: Chicago University Press, 1994).

A.N. Kolmogorov, *Grundbegriffe der Wahrscheinlichkeitsrechnung*, Ergehisse der Mathematik 2, 196-262 (1933); translated by N. Morrison as *Foundations of the Theory of Probability* (New York: Chelsea Publishing Co., 1950).

L. Krüger, L. Daston, and M. Heidelberger, eds., *The Probabilistic Revolution, Vol. 1: Ideas in History* (Cambridge, Mass.: M.I.T. Press, 1987).

L. Krüger, G. Gigerenzer, and M. Morgan, eds., *The Probabilistic Revolution, Vol. 2: Ideas in the Sciences* (Cambridge, Mass.: M.I.T. Press, 1987).

T. Kuhn, *The Structure of Scientific Revolutions* (Chicago: U.C.P., 1962; second ed. 1970).

D.A. Lane, "Fisher, Jeffreys, and the nature of probability," in *R.A. Fisher: An Appreciation* (New York: Springer-Verlag, 1980).

E.R. Lapwood, "Contributions of Sir Harold Jeffreys to theoretical geophysics," Math. Scientist 7, 69-84 (1982).

L. Laudan, *Science and Relativism* (Chicago: U.C.P., 1990).

D.V. Lindley, "Jeffreys's contribution to modern statistical thought," in A. Zellner, ed., *Bayesian Analysis in Econometrics and Statistics* (Amsterdam: North Holland, 1980).

D.V. Lindley, "On re-reading Jeffreys," in Francis, ed., *Proceedings of the Pacific Statistical Congress* (North-Holland, 1986).

D.V. Lindley, in "Fisher: a retrospective," Chance **3**, 31-2 (1990).

D.V. Lindley, "Sir Harold Jeffreys," Chance **4**, 10-14, 21 (1991).

T. Loredo, "From Laplace to supernova Sn 1987A: Bayesian inference in astrophysics," in P. Fougère, ed., *Maximum Entropy and Bayesian Methods* (Dordrecht: Kluwer, 1990).

D. MacKenzie, *Statistics in Britain, 1865-1930* (Edinburgh: Edinburgh University Press, 1981).

H.W. Macrosty, "Memorial on earnings and cost of living," Jou. Roy. Stat. Soc. **99**, 360-3 (1936); reply from T.W. Phillips, 556-7; response from Macrosty, 557-8.

W.F. Maunder, "Sir Arthur Lyon Bowley (1869-1957)," in Kendall and Plackett, 1977.

D. Mayo, *Error and the Growth of Experimental Knowledge* (Chicago: C.U.P., 1996).

E. Mayr and W.B. Provine, *The Evolutionary Synthesis: Perspectives on the Unification of Biology* (Cambridge, MA: Harvard University Press, 1980; paperback edition 1998).

J. Mehra and H. Rechenberg, *The Historical Development of Quantum Mechanics* (New York: Springer, 1982-1987).

C. Ménard, "Why was there no probabilistic revolution in economic thought," in Krüger, Gigerenzer, and Morgan, 1987.

R. Merton, *Sociology of Science* (Chicago: U.C.P., 1983).

K.H. Metz, "Paupers and numbers: the statistical argument for social reform in Britain during the period of industrialization," in L. Krüger, L. Daston, and M. Heidelberger, 1987.

J.S. Mill, *A System of Logic* (London: John W. Parker, 1843).

R. von Mises, *Probability, Statistics, and Truth* (New York: Allen and Unwin, trans. 1957).

R. Monk, *Ludwig Wittgenstein: the Duty of Genius* (London: Cape, 1990).

R. Monk, *Bertrand Russell: the Spirit of Solitude* (London: Cape, 1996).

M. Morgan, "The probabilistic revolution in economics – an overview," and "Statistics without probability and Haavelmo's revolution in econometrics," in Krüger, Gigerenzer, and Morgan, 1987.

J. von Neumann, *Mathematical Foundations of Quantum Mechanics*, trans. R.T. Berger (Princeton, N.J.: Princeton University Press, 1955).

J. Neyman, "On the two different aspects of the representative method: the method of stratified sampling and the method of purposive selection," Jou. Roy. Stat. Soc. A **97**, 558-625 (1934).

J. Neyman, "Statistical problems in agricultural experimentation," Jou. Roy. Stat. Soc. B Suppl. **2**, 107-180 (1935).

J. Neyman and E.S. Pearson, "On the use and interpretation of certain test criteria," Biometrika **20A**, 175-240 (1928).

J. Neyman and E.S. Pearson, "The testing of statistical hypotheses in relation to probabilities *a priori*," Cam. Phil. Soc. **29**, 492-510 (1933).

J. Neyman, *Lectures and Conferences on Mathematical Statistics* (Washington, D.C.: Graduate School of the United States Department of Agriculture, 1938).

N. Nicolson, ed., *The Diaries of Virginia Woolf, Volume II: 1912-1922* (New York: Harcourt Brace Jovanovich, 1976).

B.J. Norton, "Karl Pearson and statistics: the social origins of scientific innovation," Soc. Stud. Sci. **8**, 3-34 (1978).

A. Obserchall, "The two empirical roots of social theory and the probability revolution," in Krüger, Gigerenzer, and Morgan, 1987.

K.M. Olesko, "The meaning of precision: the exact sensibility in early nineteenth-century Germany," in Wise, 1995.

I. Olkin, "A conversation with Maurice Bartlett," Stat. Sci. **4**, 151-163 (1989).

D.B. Owen, ed., *On the History of Statistics and Probability* (New York: Marcel Dekker, 1976).

E.S. Pearson, "Statistical method in the control and standardization of the quality of manufactured products," Jou. Roy. Stat. Soc. **96**, 21-75 (1933).

E.S. Pearson, "Some incidents in the early history of biometry, 1890-94," Biometrika **52**, 3-18 (1965); reprinted in Pearson and Kendall, 1970.

E.S. Pearson, "The Neyman–Pearson story: 1926-34," in *Festschrift for J. Neyman* (New York: John Wiley, 1966); reprinted in Pearson and Kendall, 1970.

E.S. Pearson, "Some early correspondence between W.S. Gosset, R.A. Fisher, and Karl Pearson, with notes and comments," Biometrika **55**, 445-57 (1968); reprinted in Pearson and Kendall, 1970.

E.S. Pearson, "Memories of the impact of Fisher's work in the 1920s," Int. Stat. Rev. **42**, 5-8 (1974).

E.S. Pearson and M.G. Kendall, eds, *Studies in the History of Statistics and Probability* (London: Charles Griffin, 1970).

E.S. Pearson, eds. R.L. Plackett and G.A. Barnard, *'Student:' a Statistical Biography of William Sealy Gosset* (Oxford: O.U.P., 1990).

K. Pearson, *The Grammar of Science* (London: Walter Scott, 1892; second edition, 1900; third edition, London: Adam and Charles Black, 1911).

K. Pearson, "The fundamental problem of practical statistics," Biometrika **13**, 1-16 (1920a).

K. Pearson, "Notes on the history of correlation," Biometrika **13**, 25-45 (1920b); reprinted in Pearson and Kendall, 1970.

K. Pearson, "Further note on the χ^2 goodness of fit," Biometrika **14**, 418 (1922).

K. Pearson, *The History of Statistics in the 17th and 18th Centuries*, ed. E.S. Pearson (London: Charles Griffin, 1978).

D. Pestre and J. Krige, "Some thoughts on the early history of CERN," in P. Galison and B. Hevly, *Big Science: The Growth of Large-Scale Research* (Stanford, Calif.: Stanford University Press, 1992).

A. Pickering, "Editing and epistemology: three accounts of the discovery of the weak neutral current," Knowledge and Society **8**, 217-32 (1989).

248

A. Pickering and A. Stephanides, "Constructing quaternions: on the analysis of conceptual practice," in A. Pickering, ed., *Science as Practice and Culture* (Chicago: U.C.P., 1992).

R.L. Plackett, "Royal Statistical Society, the last 50 years: 1934-1984," Jou. Roy. Stat. Soc. A **147**, 140-50 (1984).

R.L. Plackett, Reply to S.L. Zabell, Stat. Sci. **4**, 256-258 (1989).

J. von Plato, *Creating Modern Probability: Its Mathematics, Physics and Philosophy in Historical Perspective* (Cambridge: C.U.P., 1994).

H. Poincaré, *Science and Hypothesis* (London: Walter Scott, 1905; New York: Dover, 1952).

T. Porter, *The Rise of Statistical Thinking* (Princeton Univ. Press, 1986).

T. Porter, *Trust in Numbers: the Pursuit of Objectivity in Science and Public Life* (Princeton, N.J.: Princeton Univ. Press, 1995).

W.B. Provine, *The Origins of Theoretical Population Genetics* (Chicago: University of Chicago Press, 1971).

S. Press, *Bayesian Statistics: Principles, Models, and Applications* (New York: Wiley, 1989).

F.P. Ramsey, *The Foundations of Mathematics and Other Logical Essays*, ed. R.B. Braithwaite (London: Routledge and Kegan Paul, 1931).

M.L.G. Redhead, "A Bayesian reconstruction of the methodology of scientific research programmes," Stud. Hist. Phil. Sci. **11**, 341-347 (1980).

C. Reid, *Neyman – From Life* (Springer: New York, 1982).

E.C. Rhodes, "The construction of an index of business activity," Jou. Roy. Stat. Soc. **100**, 18-66 (1937).

D. Ross, *The Origins of American Social Science* (Cambridge: C.U.P., 1991).

A. Rusnock, "Quantification, precision, and accuracy: determination of population in the Ancien Régime," in Wise, 1995.

L.J. Savage, *The Foundations of Statistics* (New York: Wiley, 1954; Dover 1972).

L.J. Savage, "On re-reading R.A. Fisher," (with discussion) Ann. Statist. **3**, 441-500 (1976).

L.J. Savage, "Reading note, 1960," in A.I. Dale, *A History of Inverse Probability: From Thomas Bayes to Karl Pearson* (New York: Springer-Verlag, 1991).

S. Schaffer, "Accurate measurement is an English science," in Wise, 1995.

S. Shapin and S. Schaffer, *Leviathan and the Air-pump: Hobbes, Boyle, and the Experimental Life* (Princeton, N.J.: Princeton University Press, 1985).

S. Shapin, *A Social History of Truth: Civility and Science in Seventeenth-century England* (Chicago: University of Chicago Press, 1994).

W.A. Shewart, *Economic Control of Quality of Manufactured Products* (New York: Van Nostrand, 1931).

D.S. Sivia, *Data Analysis: A Bayesian Tutorial* (Oxford: O.U.P., 1996).

C.W. Smith and M.N. Wise, *Energy and Empire: a Biographical Study of Lord Kelvin* (Cambridge: C.U.P., 1989).

A. Sokal and J. Bricmont, *Intellectual Impostures* (London: Profile, 1998).

H.E. Soper, A.W. Young, B.M. Cave, A. Lee, and K. Pearson, "A cooperative study. On the distribution of the correlation coefficient in small samples. Appendix to the papers of 'Student' and R.A. Fisher," Biometrika **11**, 328-413 (1917).

H.E. Soper, "Sampling moments of *n* units each drawn from an unchanging sampled population, for the point of view of semi-invariants," Jou. Roy. Stat. Soc. **93**, 104-114 (1930).

T.D. Sterling, "Publication decisions and their possible effects on inferences drawn from tests of significance – or vice versa," Jou. Am. Stat. Soc. **54**, 30-4 (1959).

S. Stigler, *History of Statistics* (Harvard Univ. Press, 1986).

J. Strachey and A. Strachey, in P. Meisel and W. Kendrick, eds., *Bloomsbury/Freud, The Letters of James and Alix Strachey 1924-1925* (New York: Basic Books, 1986).

'Student' [i.e., W.S. Gosset], "The probable error of a mean," Biometrika **6**, 1-25 (1908).

Z. Switjink, "The objectification of observation: measurement and statistical methods in the nineteenth century," in L. Krüger, L. Daston, and M. Heidelberger, 1987.

E.H. Synge, "Uniform motion in the aether," Nature **107**, 747; **108**, 80 (1921).

I. Todhunter, *History of the Mathematical Theory of Probability* (London: Macmillan, 1865).

S. Traweek, *Beamtimes and Lifetimes* (Cambridge: Harvard University Press, 1988).

J.R.G. Turner, "Random genetic drift, R.A. Fisher, and the Oxford School of ecological genetics," in Krüger, Gigerenzer, and Morgan 1987.

J. Venn, *The Logic of Chance* (London: Macmillan, 1866; New York: Chelsea, 1962).

A. Wald, *Sequential Analysis* (New York: Wiley, 1947).

A. Warwick, "Cambridge mathematics and Cavendish physics: Cunningham, Campbell and Einstein's relativity 1905-1911, Part I: The uses of theory," Stud. Hist. Phil. Sci. **23**, 625-56 (1992).

A. Warwick, "Cambridge mathematics and Cavendish physics: Cunningham, Campbell and Einstein's relativity 1905-1911, Part II: Comparing traditions in Cambridge physics," Stud. Hist. Phil. Sci. **24**, 1-25 (1993).

J. Weber, *Historical Aspects of the Bayesian Controversy* (Tucson, Arizona: Division of Economic and Business Research, 1973).

R.L. Weber, *Random Walks in Science* (London: Institute of Physics, 1973).

E.B. Wilson, "Boole's challenge problem," Jou. Am. Stat. Soc. **29**, 301-4 (1934).

M.N. Wise, ed., *The Values of Precision* (Princeton, N.J.: Princeton Univ. Press, 1995).

J. Wishart, "Statistics in agricultural research," Jou. Roy. Stat. Soc. B Suppl. **1**, 26-61 (1934).

J. Wishart, "Some aspects of the teaching of statistics," Jou. Roy. Stat. Soc. **103**, 532-564 (1939).

J. Wisniewski, "A note on inverse probability," Jou. Roy. Stat. Soc. **100**, 418-20 (1937).

L. Wittgenstein, *Tractatus Logico-Philosophicus*, trans. C.K. Ogden and F.P. Ramsey (London: Routledge, 1922).

S. Wright, "Evolution in Mendelian populations," Genetics **16**, 97-159 (1931).

D. Wrinch and H. Jeffreys, "On some aspects of the theory of probability," Phil. Mag. **38**, 715-731 (1919).

D. Wrinch and H. Jeffreys, "On certain fundamental principles of scientific inquiry," Phil. Mag. **42**, 369-90 (1921a).

D. Wrinch and H. Jeffreys, "The relationship between geometry and Einstein's theory of gravitation," Nature **106**, 806-809 (1921b).

D. Wrinch and H. Jeffreys, "On the seismic waves from the Oppau explosion of 1921 Sept. 21," Mon. Not. Roy. Astron. Soc. Geophys. Suppl. **1**, 15-22 (1923a).

D. Wrinch and H. Jeffreys, "On certain fundamental principles of scientific inquiry (second paper)," Phil. Mag. **45**, 368-374 (1923b).

D. Wrinch and H. Jeffreys, "The theory of mensuration," Phil. Mag. **46**, 1-22 (1923c).

F. Yates, "The influence of 'Statistical Methods for Research Workers' on the development of the science of statistics," Jou. Am. Stat. Ass. **46**, 19-34 (1951).

F. Yates and K. Mather, "Ronald Aylmer Fisher," Biog. Mem. Fell. Roy. Soc. **9**, 91-129 (1963).

G.U. Yule and L.N.G. Filon, "Karl Pearson, 1857-1936," Obit. Not. Fell. Roy. Soc. **2**, 73-110 (1936).

S.L. Zabell, "W.E. Johnson's 'Sufficientness' Postulate," Annals of Statistics **10**, 1091-99 (1982).

S.L. Zabell, "The Rule of Succession," Erkenntnis **31**, 283-321 (1989a).

S.L. Zabell, "R.A. Fisher on the history of Inverse Probability," Stat. Sci. **4**, 247-263 (1989b).

S.L. Zabell, "Ramsey, truth, and probability," Theoria **57**, 211-238 (1991).

S.L. Zabell, "R.A. Fisher and the fiducial argument," Stat. Sci. **7**, 369-387 (1992).

A. Zellner, "Introduction," in A. Zellner, ed., *Bayesian Analysis in Econometrics and Statistics* (Amsterdam: North Holland, 1980).

Index

actuarial science, 47, 51, 174n
Adams, J.C., 236
agriculture, 8, 61, 70–2, 73, 172–3, 176, 198
Aldrich, John, 67n, 68, 80n
Alembert, Jean d', 31, 38n
Amador, Risueño d', 38
annuities, 11, 15–6, 17–9
Arbuthnot, John, 33, 213
Aristotelianism, 16n
Arnauld, Antoine, 17n
astronomy, 4, 11, 16, 19, 24, 25, 28, 35, 74,
     111–3, 116, 116n, 132, 172, 207–8
astrophysics, *see* solar system
atmospheric physics, 4, 83–4, 113n
axioms, 90–1, 90n, 100, 108, 115, 163, 216,
     219–20, 226
Ayer, A.J., 87n

Ball, Sir Robert, 81
balls-in-urn, *see* urn model
Barnard, George, 61, 68, 138, 180n, 183,
     190n, 223n
Bartlett, M.S., 124n, 139–41, 143, 152, 161–2,
     162n, 164–5, 168, 179n, 180, 184n,
     188n, 190–1, 224
Bateson, William, 54–6
Bayes, Thomas, 1, 2, 23–4, 25, 37n, 65n, 68n,
     90n, 114, 154n, 164, 164n
Bayes's theorem, 30–1, 49–50, 59n, 68, 69,
     80, 90, 100, 104, 107, 118, 122, 127,
     131, 144, 146, 148, 158, 171, 183, 218,
     218n, 225, 235–6
Bayes's postulate, *see* Principle of Insufficient
     Reason
Bayesian interpretation, *see* probability,
     inverse
Becquerel, Henri, 203
behaviorism, 194n
Behrens, W.U., 180n
Behrens–Fisher problem, 180, 188n, 230
belief, degrees of, *see also* probability, inverse,
     1, 3, 11, 27, 30, 32n, 36, 37–8, 48–9,

62, 64n, 68n, 88, 89, 97n, 98, 114, 119,
     123, 139, 140, 148, 151–3, 158, 163,
     191, 196, 209, 211, 217
Bernard, Claude, 39, 44
Bernoulli, Daniel, 21, 25, 29, 38, 63n
Bernoulli, Jacob, 15, 19–20, 20n, 22, 23, 25n,
     27, 37
Bernoulli, Nicholas, 15, 20n, 21, 21n, 22, 29
Bertillon, Alphonse, 44
Bertrand, Alexis, 48n
Bessel, Frederick, 47
Bibby, John, 186n
billiard table, 37n, 65n
bills of mortality, *see* mortality, bills of
binomial distribution, *see* distribution,
     binomial
biology and biologists, 42, 172, 213–6
biometrics, 4, 11, 44–7, 53–6, 60n, 70, 126,
     134, 172, 186, 213–4
birth ratio and birth rate, 25, 34, 185, 213
Black, Dora, 109
black-body radiation, 203
Blackett, P.M.S., 121
Bletchley Park, 223n, 226
blocking, 198
Bobeck, Professor, 89n, 99
Bohr, Niels, 204, 204n, 205, 206n, 210, 215
Boltzmann, Ludwig, 42, 48, 62n, 200–2, 203,
     204
Bolzano, Bernard, 32n
Boole, George, 32, 38, 49, 50n, 80, 88, 99,
     171n
Boole's Challenge Problem, 189n
Borel, Emile, 202n, 216
Boring, Edwin, 194n
Born, Max, 205
Bortkiewicz, L. von, 36–7n, 99n
Boscovitch, Roger, 17
botany, 82, 102n, 127
Bowley, Sir Arthur, 172, 172n, 175n, 183, 186,
     186n, 187, 187n, 188, 190, 190n, 193,
     195n

253

Box, Joan Fisher, 12, 65, 68, 176n, 180n
Braithwaite, Richard, 220–1
Brandt, A.E., 173
Breuer, Joseph, 118n
Bridgman, Percy, 103n, 207
Brillouin, Marcel, 113n
Broad, C.D., 76n, 87–8, 87n, 89n, 90–2, 96–7,
        97n, 98n, 99n, 104–5, 106, 106n, 119,
        123, 125, 140n, 144, 149, 220, 200n
Bromwich, T.J. d'A., 113n
Broussais, F.J.V., 38n
Brouwer, L.E.J., 206
Brown, Ernest, 110
Brown, Robert, 203
Brownian motion, 62, 202–3, 204
Brownlee, John, 172
Brunswick, Egon, 194n
Buckle, Henry, 40, 41n, 48, 200, 201n
Buffon, G., 24
Bullen, Ken, 138, 222, 222n
Bulwer, Henry Lytton, 34
Bunyakovsky, Victor, 48n
bureaucracy, 6, 11, 33, 35
Byerly, Perry, 129n

calculating machines, 82n, 110, 140n
Cambridge, 7, 56, 58, 81–2, 86, 87n, 98, 109,
        139, 172–3, 179n, 223–4
Campbell, Norman, 102n, 207, 209, 210n
card games, see games of chance
Cardano, Girolamo, 14n
Carnap, Rudolph, 220, 220n
Cartwright, Nancy, 206
causality, see causes and effects
causation, see causes and effects
causes and effects, see also determinism, 3, 10,
        23, 26, 27, 30–2, 39, 71, 204–6, 212
census, 33, 48
central limit theorem, see limit theorem
Ceres (asteroid), 26
Chamberlain, Thomas, 112
Chambers, G.F., 81
chance and chances, see also randomness, 7,
        18, 27, 36, 61–2, 140, 140n, 143, 152,
        152n, 161–2, 164, 167, 190–1, 194,
        201–2, 203–4, 203n, 205–6, 209–11,
        215, 219n
Chaplin, F.S., 194
Chauvenet, William, 132
chest measurements, 35
χ^2(chi-square) test, 46, 57, 60, 65, 66–7, 67n
Chrystal, George, 51, 51n, 75, 80, 80n, 171n
Church, Alonzo, 217
class, social, 5, 44–6, 52, 55–6, 58n
Clausius, Rudolf, 199, 200
clinical trials, 39n, 172, 182n
Cochran, W.G., 173n, 189, 189n

coin tossing, 1–2, 21, 32, 62, 72n, 211, 211n
combination of observations, 11, 16–7, 25
combinatorial analysis, 14, 15, 31, 62, 69
common sense, 5, 49, 89, 107, 108, 117, 119,
        120, 124, 124n, 127, 162, 182, 186
complementarity, 205, 215
componential analysis, 57
Comte, Augustus, 39, 40n
Condon, E.U., 207
Condorcet, Antoine, Marquis de, 22, 24, 25,
        32, 33
confidence levels, see confidence limits
confidence limits, 47, 175, 177–8, 188n, 228,
        229
confidence internals, see confidence limits
Conniffe, D, 68, 99n
Connor, L.R., 180n, 182, 189
conscription, 36
consistent statistic, 66, 66n
control groups, 71
Cook, Alan, 110n, 138n
Coolidge, Julian, 171
Copeland, Arthur, 219
Copenhagen interpretation, 205, 206–7, 210
correlation, 11, 44, 50, 58, 76n, 193–4
        coefficient, 57, 59, 63, 79n
Correns, Carl, 54
Cotes, Roger, 16, 16n
Cournot, Antoine, 37, 38
crab data, 46, 54
Craig, John, 22
Cramér, Harald, 189, 190n
credibility, 6, 10, 34, 40n, 74, 185, 191–2
crime, 3, 33, 35–6, 40–1
Crofton, Morgan, 51
crows, 87–8, 88n, 90–2, 104–5, 124–5
cryptanalysis, 223n, 226
Cunningham, Ebenezer, 82
Curie, Pierre, 84
curve fitting, see data analysis
Cuvier, Georges, 44

Dalton, John, 20
Daniels, H.E., 180n
Darwin, Charles, 42, 43, 44, 53, 62n, 84
Darwin, Sir George, 81, 117
Darwin, Leonard, 58, 60, 62, 66
Daston, Lorraine, 30n, 41, 192n, 231
data analysis, 10, 11, 25, 76, 105–7, 128–9,
        133–4, 159, 160, 169, 192–3, 196, 198,
        222
data reduction, see data analysis
David, F.N., 180, 232
decision making, 175, 178–9, 225, 226, 229
decision theory, see also decision making, 10,
        223
degree of belief, see belief, degrees of

degrees of freedom, 67, 208n
Descartes, René, 27
design of experiments, *see* experimental design
determinism, *see also* causes and effects, 2, 9,
    27–8, 40, 42, 61, 121n, 193–4, 194n,
    199–202, 200n, 202n, 204, 205–6,
    206n, 207–9, 209–13, 215–6, 217,
    219n, 220n, 227–8
deviation
    social, *see also* crime, 35, 40
    statistical, 44, 136
    standard, 56, 57n, 58n, 60, 76, 79n, 130n,
        132–3, 132n, 133n, 141, 146, 155,
        162n, 213n
dice games, *see* games of chance
Dingle, Herbert, 208–9
Dirac, Paul, 205, 209
disease, 3, 15, 35
distribution,
    asymmetric, 46, 46n, 121–6
    bell-shaped, *see* normal
    binomial, 23, 25n, 36n
    error, 17, 24, 25, 26, 28–9, 72n, 129–33,
        135–6, 146n
    Gaussian, *see* normal
    normal, 25, 26, 28–9, 35, 43–4, 46, 130,
        133n, 142, 146, 174, 199
    sampling, *see also* sampling, 63, 65, 72–3,
        77, 121–6
    t-, 57, 76
    z-, *see* t-
Dobzhansky, Theodosius, 213n
doctrine of chances, 3, 19–22, 27
Donkin, William, 49
Doob, J.L., 219
Double, François, 38–9
dreams, 117–8, 118n
Drinkwater-Bethune, John, 27
Dudding, B.P., 180n, 182
Durkheim, Emile, 48

Earth, 17, 82–3, 84, 94–6, 111–2, 113n, 116n,
    126, 128, 222
    age of, 53–4, 84
    core, 4, 8, 106, 112
earthquakes, *see* seismology
eclipse, 84, 92–3, 207
econometrics, 186, 186n, 193n, 225
economics and economists, 97n–98n, 98,
    172n, 181, 187, 191n, 193, 225
Eddington, Sir Arthur, 10, 82, 89, 92–3, 104,
    106, 162n, 163, 166, 172, 207–8, 207n,
    208n, 211
Edgeworth, Francis Ysidoro, 43, 44n, 46n, 47,
    50, 58n, 59, 59n, 80, 183, 183n, 193,
    193n
Edinburgh school, 9

Edwards, A.W.F., 63n, 68
Edwards, Ward, 224–5
efficient statistic, 66, 66n
Einstein, Albert, 7n, 10, 85, 92–3, 103, 116,
    163, 203–4, 203n
Eimer, Theodore, 53
Elderton, Sir William, 187n
Ellis, Richard Leslie, 32, 37, 40
Elsasser, Walter, 96
embryology, 54
ensemble, 8, 37, 41, 48, 139, 155, 204, 210
entropy, 202
enumeration, *see* induction, *also* urn model
epistemic interpretation, *see* probability,
    inverse
equally-possible cases, *see also* Principle of
    Insufficient Reason, 15, 20, 30
Equitable Company, 19, 19n
ergodic theory, 205n
error
    experimental, 16–7, 24, 106, 128–33, 135–6
    first and second kind, 174–5, 178, 223
    theory, 11, 16–7, 25–6, 28–9, 35, 36, 47, 52,
        63n, 74, 89, 172
    estimation, 72n, 98n, 159–60, 164, 174
    tests, 67, 71n, 196n
eugenics, 8, 44–6, 52–3, 55–6, 56n, 58, 58n,
    67, 185n, 192, 192n
Euclid, 94, 104
Euler, Leonard, 18
evidence, legal, 22, 26, 28, 32
evolution, 4, 41–7, 53–6, 84, 213–6
    social, 40
evolutionary synthesis, 4, 214–5
exact sensibility, *see* precision, culture of
exchangeability, 220n
Exner, Franz, 204, 206n
expectation, mathematical, 2, 15, 17, 21–2,
    21n, 68n, 154n, 164
experiment, 127, 163, 194, 198, 209, 212
    genetic, 58, 215
    repeatable, 70, 168–9, 177–8, 217
experimental design, 8, 70–3, 192, 196, 198
expertise, professional, 8, 19, 28, 39, 132, 162,
    179, 191–2, 196, 230
extrapolation, 104–6, 107–8, 113, 126, 156–7

falsifiability, 49, 221
farming, 58
Farrow, E.P., 82, 86, 86n
Fechner, Gustav, 193, 194n, 217n
Fermat, Pierre de, 2, 14–5, 14n, 21
fiducial probability, *see* probability, fiducial
Fieller, E.C., 180n
Filon, L.N.G., 50
Finetti, Bruno de, 3, 11, 13, 212n, 220n,
    227–8, 227n, 229n

Finney, D.J., 138n
Fisher, Sir Ronald, 4, 7–13, 37n, 89, 119, 121,
    126–7, 173–6, 181–2, 182n, 191n,
    183–6, 186n, 188, 188n, 189, 195n,
    196, 196n, 198, 217–8, 219, 221, 226,
    228–30
  background and education, 52–3, 56
  debate with Jeffreys, 4, 9, 12, 128–70, 176n
  early work & career, 56–9, 75, 137
  and eugenics, 56, 58, 58n, 67, 185n
  and evolutionary synthesis, 4, 213–6
  and fiducial probability, 68n, 75–9, 134,
      136, 138, 155–6, 162, 168, 175, 175n,
      177, 180, 188n, 229–30, 230, 230n
  and frequency definition, 63–5, 69–75, 78n,
      140–3, 151–2, 155–6, 158–9
  and genetics, 7, 56, 61, 73, 75, 122, 133,
      167, 169, 192, 213–5
  and Neyman, 175–80, 184n, 186n, 189,
      229–30
  parachute problems, 168
  and Pearson, 59–61, 65–70, 80, 134
  and Principle of Insufficient Reason, 62,
      64–5, 65n, 68, 79, 80, 148–50, 152–3
  and inverse probability, 61–70, 75, 78–80,
      100n, 122–3, 128, 134–6, 152–4, 171,
      188n
  and Rule of Succession, 80, 100n
  and statistical inference, 79–81, 155–6,
      159–62, 164, 178–9, 179n, 183–6,
      184n
  Statistical Methods for Research Workers,
      72–6, 134, 138, 184, 190, 192, 195, 215
  style and personality, 138, 165–6, 180
Fisherian statistics, 12, 191–8, 198n, 215,
    223–4
fluctuation, 23, 202–3
fluxions, 23
Forbes, J.D., 50n
Forman, Paul, 206, 206n, 207
Fourier, Joseph, 47
Fréchet, M., 190
free will, *see also* determinism, 40, 41n, 42,
    42n, 201n
Freeman, Austin, 126n
French Revolution, 22, 29, 32, 38
frequency interpretation, *see* probability,
    frequency interpretation
Freud, Sigmund, 86n, 102n, 117, 118n
Fries, Jacob, 37, 38, 88n

Galileo, Galilei, 14n, 16, 16n
Galison, P., 192n
Galton, Francis, 43–44, 46n, 50, 54, 60, 63,
    76n, 98, 168
Galton Laboratory, 60, 67n, 137, 172, 173,
    174, 175

gambling, *see* games of chance
games of chance, 2, 11, 14–5, 17, 18, 19–20,
    27, 28, 31, 74, 97, 152–3, 182n, 217, 231
Garrod, Sir Archibald, 54
Gauss, Karl, 25–6, 28–9, 31n, 63n, 65, 130n
Geiger, Hans, 203
Geisser, S., 224, 228n
gene frequency, 4, 69, 214
genetic drift, 214–5, 215n
genetics, 7, 54–6, 58, 61, 73, 75, 121–2, 133,
    161, 167, 169, 213–5
gentlemanly science, 6, 43
geology, 53n, 81, 111–2, 127
geophysics, 4, 111–2, 126–7, 133, 161, 169,
    222
Gibbs, J.W., 57, 204
Giddings, Franklin, 194
Gigerenzer, Gerd, 194, 194n, 197, 197n, 198n
Glenday, R., 185, 186n
God and God's will, 18, 20, 27, 34
Goldschmidt, Richard, 214
Good, I.J., 3, 97n, 131n, 224, 225, 226, 228,
    228n
Gooday, G., 197
Gordon, A.P.L., 182n
Goschen, Lord, 67n
Gosset, W.S., 50, 56–7, 57n, 59, 68n, 69, 71,
    74, 76n, 77, 138, 174, 180, 180n, 191n,
    195, 213n
Goulden, C.H., 189
Govan, John, 51
Graunt, John, 15, 34
gravity, 17, 85, 92–3, 109, 112, 115, 116, 120,
    157, 193
Greenwood, Major, 172, 182, 183, 185, 186
Grummell, E.S., 180n
Guinness Brewery, 57
Guinness, Henry, 58
Guinness, Ruth, 58
Gutenberg, B., 112, 222

Haavelmo, Trygve, 193
Hacking, Ian, 20, 33, 38n, 41, 200, 231
Haldane, J.B.S., 4, 121–6, 131, 131n, 133–4,
    137, 144–5, 157, 215, 217n
Hall, Asaph, 108
Halley, Edmund, 16
Hamilton, William, 7n
Hardy, G.H., 223n
Hardy, George, 51
Hartley, David, 29
Hartley, H.O., 173n, 179n
Harrison, J.W.H., 82
Heaviside, Oliver, 113, 113n
Heisenberg, Werner, 205, 206, 210, 212
heredity, 4, 11, 42–5, 52–6, 57–8, 64, 169,
    192n, 213–4, 217

Herschel, Sir John, 50, 199
heterozygosity, 63
hidden variables, 210
Hilbert, David, 216
Hill, A.B., 182n
Hippocrates, 39
historiography of science, 5–7, 12, 116n, 206
Holmes, Arthur, 84
homozygosity, 50
Hooke, Robert, 183n
Hooper, George, 22
horses, 27, 36n–37n
Hotelling, Harold, 74, 195
H-theory, 200–1, 202
Huilier, Simon L', 88n
Hull, Clark, 194
Hume, David, 16, 20, 23, 24, 27, 29, 29n, 30n
Huxley, Thomas, 42
Huygens, Christian, 15–6
Huzurbazar, V., 224
hydrodynamics, 83, 216

Ibsen, H., 117
ignorance, *see also* knowledge and Principle of Insufficient Reason, 27, 31–2, 37–8, 64, 145–7, 160, 230
indeterminism, *see* determinism
induction, *see also* inference, 23–4, 48, 49, 72, 75, 96, 97n, 101n, 102, 107, 122–6, 133, 226
inference, 8, 10, 37, 47, 50, 68, 76, 86–8, 95–6, 100, 103–4, 109, 114–9, 122–6, 149–50, 151–2, 153
    statistical, 70–1, 79–80, 155–6, 161–2, 164, 176, 178–9, 183–6, 187–8, 190–1, 194, 196–7, 224, 230
inheritance, *see* heredity
inoculation, 38, 38n
insufficient reason, principle, *see* Principle of Insufficient Reason
insurance, 15–6, 18–9, 18n, 27, 28, 37, 39, 51, 99, 231
intuition, 18, 107, 119, 123–4, 124n, 146n, 225
inverse probability, *see* probability, inverse
irrationality, *see* rationality
irreversibility, 42, 202
Irwin, Oscar, 173, 173n, 189, 191n
Isserlis, Leon, 183, 183n, 188, 188n

Jansenism, 17n
Jaynes, E.T., 225
Jeans, James, 52, 84, 114
Jeffreys, Lady Bertha, 110n
Jeffreys, Sir Harold, 4, 7–13, 178n, 187, 189, 191, 196, 207, 210n, 211–2, 217, 226, 227, 227n, 228–30

and astrophysics, 4, 8, 82–3, 84–5, 111–3, 126
background and education, 81–3
career, 85, 110–3, 110n
debate with Fisher, 4, 9, 12, 128–70, 176n
and frequency theory, 89, 100, 107, 151–2, 155–6, 158–9, 163–4, 184n
and geophysics, *see* and seismology
at Meteorological Office, 83–5, 96, 109
and Principle of Insufficient Reason, 98–100, 148–50, 165
and prior probabilities, 119–26, 130–3, 135–6, 137, 140, 142–3, 144–7, 157–8, 165, 167–8
and *dh/h* prior, 122, 124–5, 130–3, 131n, 135–6, 137, 142, 146–7, 165, 165n
and probability, 89–92, 107–9, 126–7, 150–2, 152–4, 155–6, 163–70, 171, 222–5
and psychoanalysis, 101–3, 117–8, 119
and relativity, 92–4
Scientific Inference, 113–9, 145n, 196n, 211–2, 217n
and seismology, 4, 94–6, 104, 111–2, 126–7, 128–9, 132–3, 156–7, 161, 169, 222
and Simplicity Postulate, 103–7, 108, 108n, 114–5, 118, 119–21, 129, 145n, 158, 229, 229n
Theory of Probability, 166n, 189, 196n, 222, 222n, 225, 228, 228n
tramcar problem, 167–8
work style and personality, 12, 82–3, 84, 110, 127, 138–9, 165–6,169
and Dorothy Wrinch, 85–6, 94–6, 103–9
Jelliffe, S.E., 117
Jenner, Edward, 38
Jennett, W.J., 180n
Jevons, William Stanley, 1, 47, 49, 91, 98–9, 193
Johnson, W.E., 87, 97, 98, 114n, 220, 220n
Jones, Ernest, 102n, 119
Jordan, Pascual, 205, 206n
Jordan, Wilhelm, 28n, 132
juries, 3, 22, 32, 34
justice, 29

Kammerer, Paul, 54n
Kant, Immanuel, 40
Kelvin, Lord, *see* Thomson, William
Kendall, M.G., 186n, 189, 191n, 218n, 223
Kennet, Lord, 184, 185n
Kepler, Johannes, 16
Keynes, John Maynard, 10, 52, 63n, 76n, 89n, 96–100, 105n, 114n, 118, 144, 146n, 148–50, 171, 188, 189, 193, 193n, 211, 220, 227, 227n, 232
Khintchine, A.J., 219
kinetic theory, *see also* statistical physics, 42, 199–202, 203, 204, 206n, 216

Mohorovičić, Andrija, 95–6
moments, method of, 46, 66–7
Moivre, Abraham de, 15–6, 17, 20, 22, 23, 25n, 27, 63n
Montesquieu, Baron de, 16
Montmort, Pierre de, 15, 21, 22
Moon, 17, 82, 84, 111–2, 113n, 236
moral
  certainty, 20
  imperative, 4, 28n, 74
morality, 45, 51
Morgan, Augustus de, 27, 31, 32, 49, 99
Morgan, T.H., 213–4, 214n
Morgenstern, Oskar, 226
Morley, E.W., 116, 116n
mortality
  laws, 34, 38n, 217
  tables, 15–6
Moulton, Forest, 112n
Musil, Robert, 216
mutation, 213, 214

Napoleon, 99n
National Union of Scientific Workers, 85, 102n, 207
natural philosophy, 18
natural selection, 44, 53–6, 84, 112, 213–5
natural theology, 33, 42
naturalism, 81
navigation, 17
Nernst, Walter, 206n
Neumann, John von, 210, 226
Newall, H.F., 82, 83
Newman, H.A., 229n
Newtonianism, 16, 17, 23, 27, 85–6, 92, 103, 108, 113, 114, 115–6, 120, 193
Neyman, Jerzy, 10, 13, 75n, 174–6, 176–80, 183n, 184n, 186n, 187, 188n, 189, 191n, 195n, 196, 198, 219, 223, 229–30
Neyman–Pearson theory, 10, 11, 12, 174–5, 177–9, 188, 196–7, 223, 223n, 224, 226
Nicholson, John, 110
Nicod, Jean, 220
Nietzsche, Friedrich, 43n
normal curve, see distribution, normal
notation, 114n
nova, 16n
null hypothesis, 72–3, 155, 174–5, 177–8, 198, 198n, 221
nutation, 83, 83n

objectivity and subjectivity, 8, 10, 12, 28, 30, 34, 74, 148, 150–2, 154–5, 183, 185,

191n, 191–2, 192n, 196–7, 197n, 208, 210–11, 218, 224–5, 227, 229
observation, experimental, 8, 24, 127, 129, 132–3, 168–9
Oldham, R.D., 95, 112
operationalism, 103n, 207
Oppau explosion, 94–6, 111–2, 128
Oppenheimer, J.R., 207
optimum value, 63, 65, 124n, 162
orbits, 25, 26, 31
orthogenesis, 53
outliers, 19, 132–3, 133n, 222

pangenesis, 43
Pascal, Blaise, 2, 14–5, 14n, 21, 27
Pascal's wager, 20
Pearl, R., 173
Pearson, Egon, 10, 13, 68, 137, 173, 174–6, 178, 179n, 182, 184n, 190, 232, 233
Pearson, Karl, 7, 45, 47n, 50, 52, 54–6, 76n, 79, 86, 89, 92, 93, 96, 101–2, 105, 107, 112n, 115, 117, 127, 172–3, 174, 181n, 186, 187, 190, 194, 197n, 232
  and Fisher, 57–8, 59–61, 65–70, 79–80, 137
pedigree analysis, 63, 64, 73
Peirce, Benjamin, 132
Peirce, C.S., 132
Penrose, L., 117
perihelion, 31
  of Mercury, 93, 93n, 104, 109, 116, 120
Perrin, Jean, 203
personalism, see probability, personal
Phillips, T.W., 184n
photoelectric effect, 203
photography, 81, 81n
physics, 41–2, 199–213
  and probability, 199, 202–4, 209–11, 211–3, 225
Poincaré, Henri, 51, 91, 201–2, 204
Planck, Max, 203, 204
Plato, J. von, 219, 232
Poisson, Siméon-Denis, 20, 20n, 28, 29, 32–3, 34, 36, 36n, 94, 140
political arithmetic, 16n, 33
Popper, Sir Karl, 13, 220–1, 229, 229n
population
  estimates, 15–16
  infinite hypothetical, 63, 63n, 65, 68, 72, 73, 75, 81, 122n, 139, 140, 155-6, 164, 164n, 168–9, 174, 188, 190–1, 198, 215, 217
population genetics, 214–5, 221
Porter, Theodore, 18, 41, 48n, 192n, 198n, 231–2
positivism, 35, 39, 46–7, 67, 86, 103n, 106, 115, 116n, 199, 205, 216, 220n